A Colour Atlas of Burn Injuries

CHAPMAN & HALL MEDICAL ATLAS SERIES

Chapman & Hall's new series of highly illustrated books covers a broad spectrum of topics in clinical medicine and surgery. Each title is unique in that it deals with a specific subject in an authoritative and comprehensive manner.

All titles in the series are up to date and feature substantial amounts of top quality illustrative material, combining colour and black and white photographs and often specially-developed line artwork.

The amount of supporting text varies: where the illustrations are backed-up by large amounts of integrated text the volume has been called 'A text and atlas' to indicate that it can be used not only as a high quality colour reference source but also as a textbook.

Slide Atlases are also available for some of the titles in the series.

1 **A Colour Atlas of Endovascular Surgery**
R.A. White and G.H. White

Also available:
A Slide Atlas of Endovascular Surgery

2 **A Colour Atlas of Heart Disease**
G.C. Sutton and K.M. Fox

3 **A Colour Atlas of Breast Histopathology**
M. Trojani

4 **A Text and Atlas of Strabismus Surgery**
R. Richards

Also available:
A Slide Atlas of Strabismus Surgery

5 **A Text and Atlas of Integrated Colposcopy**
M.C. Anderson, J.A. Jordon, A.R. Morse and F. Sharp

6 **A Text and Atlas of Liver Ultrasound**
H. Bismuth, F. Kunstlinger and D. Castaing

7 **A Colour Atlas of Nuclear Cardiology**
M.L. Goris and J. Bretille

8 **A Colour Atlas of Diseases of the Vulva**
C.M. Ridley, J.D. Oriel and A.J. Robinson

9 **A Colour Atlas of Burn Injuries**
J.A. Clarke

In preparation

A Text and Atlas of Paediatric Oral Medicine and Pathology
R.K. Hall

A Text and Atlas of Clinical Retinopathies
P.M. Dodson, E.E. Kritzinger and D.G. Beevers

A Colour Atlas of Medical Entomology
N.R.H. Burgess

A Colour Atlas of Retinovascular Disease
S.T.D. Roxburgh, W.M. Haining and E. Rosen

A Colour Atlas of Forensic Medicine
J.K. Mason and A. Usher

A Colour Atlas of Neonatal Pathology
D. de Sa

A Text and Atlas of Minor Surgery
Second edition
J.S. Brown

A Colour Atlas of Burn Injuries

John A. Clarke, FRCS
Regional Burns Unit
Queen Mary's Hospital
Roehampton, London

CHAPMAN & HALL MEDICAL
London · Glasgow · New York · Tokyo · Melbourne · Madras

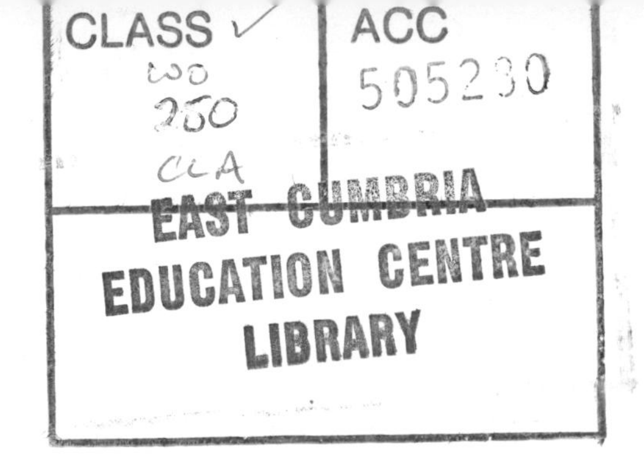

Published by Chapman & Hall, 2-6 Boundary Row, London SE1 8HN

Chapman & Hall, 2-6 Boundary Row, London SE1 8HN, UK

Blackie Academic & Professional, Wester Cleddens Road,
Bishopbriggs, Glasgow G64 2NZ, UK

Van Nostrand Reinhold Inc., 115 5th Avenue,
New York NY10003, USA

Chapman & Hall Japan, Thomson Publishing Japan, Hirakawacho
Nemoto Building, 7F, 1-7-11 Hirakawa-cho, Chiyoda-ku, Tokyo 102,
Japan

Chapman & Hall Australia, Thomas Nelson Australia,
102 Dodds Street, South Melbourne, Victoria 3205, Australia

Chapman & Hall India, R. Seshadri, 32 Second Main Road, CIT East,
Madras 600 035, India

First edition 1992

Designed by Thin Blue Line, London

Typeset in 10/12pt Palatino by Ian Foulis & Associates, Saltash
Cornwall

Printed in Hong Kong

ISBN 0 412 36520 0

A catalogue record for this book is available from the British Library

Contents

Preface

There are several excellent text-books on burns and their management, but few are well illustrated and almost none in colour. Most of these texts are written for specialists, or those planning to become specialists, or cater for special interest groups. There is therefore an opportunity for a book that is written for those who may work in a burns centre, or for others who deal with burn patients and who need a broader view than the burns specialist.

The study of burns lends itself well to visual representation as the injuries start, at least initially, in the skin and are therefore readily observable. This Colour Atlas is designed for those who wish to find out more about the different kinds of burn, how they are caused and how they can be managed. It is for those who are seeking an introduction, whether nurses, physiotherapists, occupational therapists, dieticians, pharmacists, social workers or psychologists, for doctors in training, those preparing for higher exams, for bacteriologists and pathologists.

The text is short and didactic, and is used to elaborate and explain the illustrations and their legends. I have deliberately avoided being controversial. Many areas have been glossed over or ignored, but it was felt that it was better to have a short interesting text that could be quickly comprehended, rather than an all-embracing text that would be too detailed and of less general appeal. The reader is advised to go to the textbooks listed in the bibliography for the arguments and further opinions and advice. There is a generous index and a specially selected list of excellent references which give a review of many topics, and are themselves a source of further and specific references.

The Colour Atlas is written and published as a mark of respect to the nurses in the Burns Unit at Queen Mary's Hospital, Roehampton. I sincerely admire their care, concern and love. It is a privilege to work with them. The nursing of patients with burns remains one of the few areas in medicine where individual nurses can be of profound and lasting influence. Their compassion and gentleness, their concern and understanding, is remembered for ever by patients and their relatives. Even when it is clear that one of their patients cannot survive an overwhelming injury, they are comforted and guarded, and treated with respect and dignity. I honour all who work with burn patients.

To Jonathan Chandy, who has written the section on anaesthesia, my sincere thanks. Enormously kind and considerate to his patients, who have got to know him well, he has been unstinting with his time and concern. He has overseen all patients on ventilators and those critically ill, and has, together with Alain Landes, safely cared for those undergoing surgery and repeated dressing changes. He has shared much of the burden of management of the acute patients, and has been responsible for the training of many young anaesthetists in burn care.

My gratitude is offered to all those who work in the Burns Unit at Roehampton, in particular Ruth Clarke and Debbie Steer, nursing sisters, Manu Perinpanayagam and Mike Weinbren, microbiologists, and Duro Johnson, social worker, Muriel Gaul, dietician, Claire Clarke, physiotherapist, and Sarah Wright, occupational therapist, and the many trainee plastic surgeons who have worked hard and long into the night.

I am indebted to Bob Jankowski and Neil Maffre, and others in the Department of Medical Illustration at Roehampton for the photographs. These are obviously the greatest strength and asset of an Atlas, and they have taken great pains and trouble to reflect the injuries as they really are. They have always been enthusiastic and very considerate to the patients, and without them the work could not exist. I am particularly grateful to my friends, Arieh Eldad, for the illustrations of phosphorous burns, and Ian Wilson, for the picture of mustard gas burns. All other illustrations are of patients from Roehampton, and my thanks go to them for allowing me to publish pictures of them when not looking their best.

Finally, I should like to thank the publishers for their patience and understanding, and my wife and children who were less patient but very tolerant, yet perhaps more understanding.

John Clarke
Roehampton

1 Classification of burns

INTRODUCTION

Burns are coagulative lesions involving the surface layers of the body. They are usually caused by heat, but burn injuries caused by chemical agents and radiation are also included in this description.

Clinically, classification of burns is based on the depth of tissue damage. Heat causes protein coagulation, and obviously the depth of the burn will be directly related to the temperature and the duration of contact of the causative agent.

Burns may be classified into two main groups: of partial thickness and of full thickness. Fig. 1.1 shows diagrammatically the depth of injury and pathophysiological changes that occur in different classifications of burns.

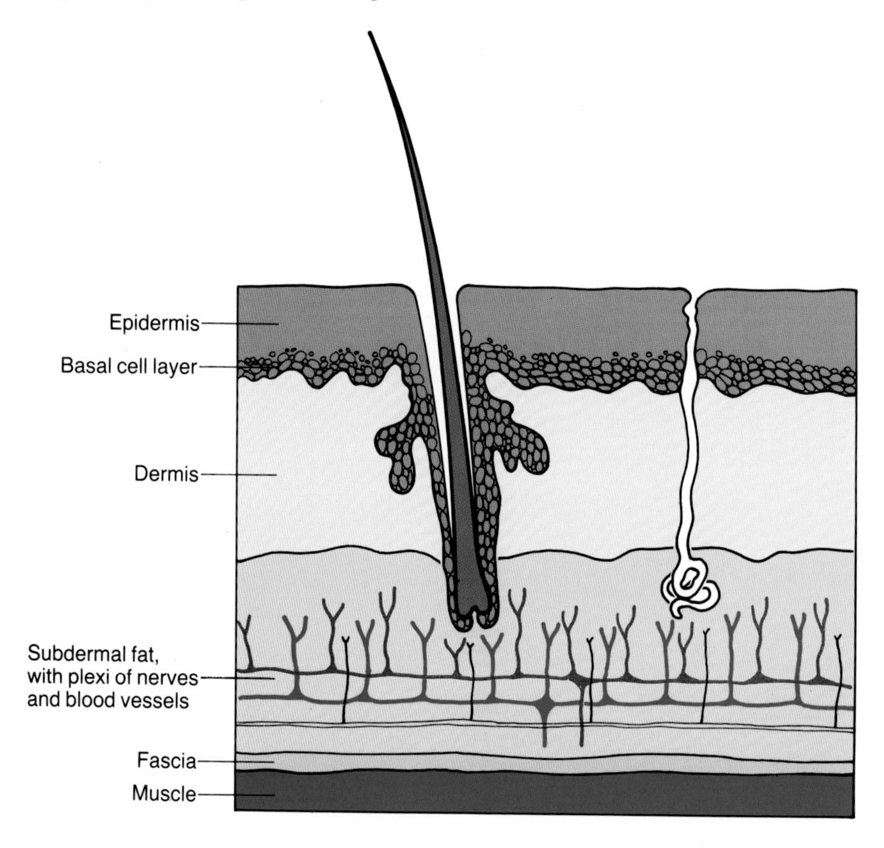

Fig. 1.1 Diagrammatic representation of the patho physiological changes which occur after burning.

(a) Cross section showing structural features. The bases of the hair follicle and sweat glands reach down into the subdermal fat and are enclosed by basal cells; these are the source of new epithelium when the dermis has been destroyed.

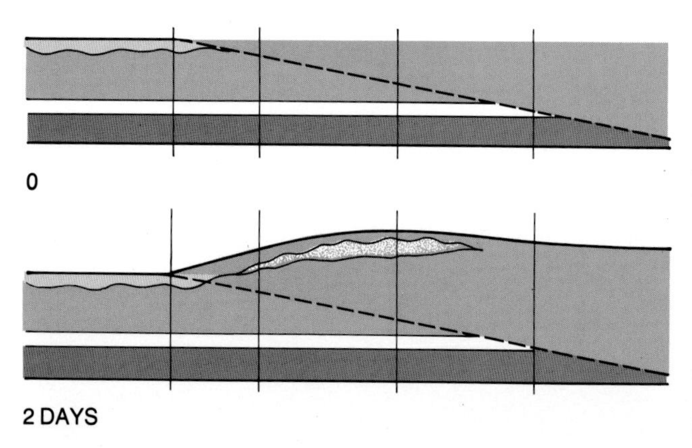

(b) Heat damaged skin giving rise to (1) Superficial, (2) Superficial dermal, (3) Deep dermal and (4) Full thickness burn.

(c) Two days after injury. Increased permeability has caused swelling. Blistering occurs in those tissues which retain a functioning vascular bed and is therefore absent in the deep burn.

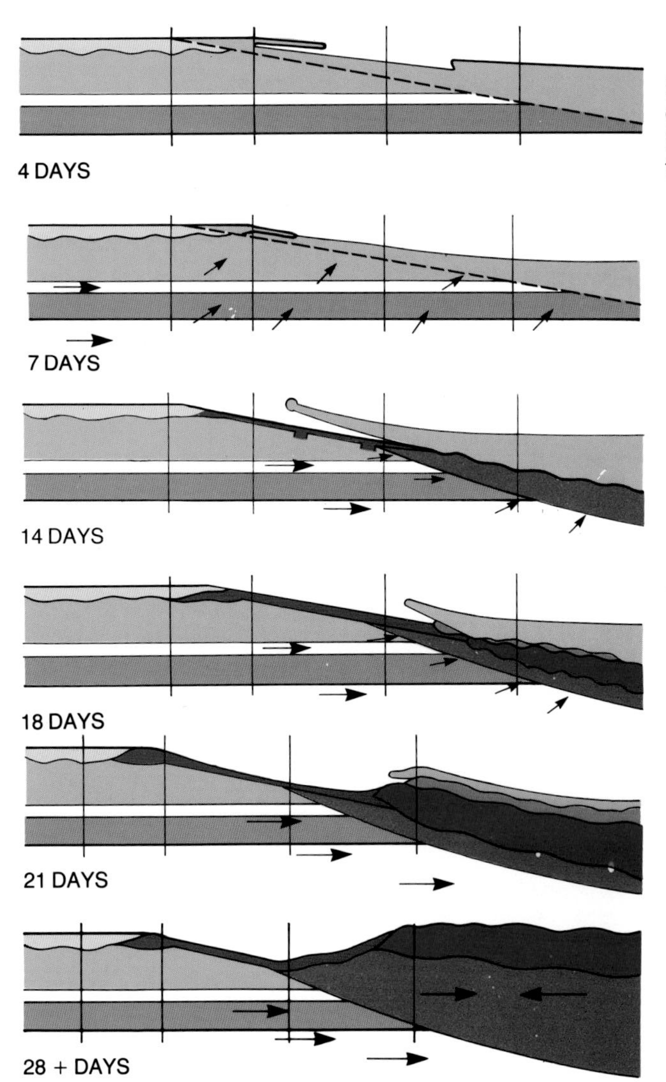

(d) Four days after injury. Absorption of fluid back into the circulation has led to dessication of the wound. The dry, inelastic eschar has sunk below the original surface contour.

(e) Seven days after injury. Blood flow in the area of the wound has increased, spontaneous regeneration of epithelium from basal cells has started.

(f) Fourteen days after injury. Basal cell regeneration and restoration of the epithelium is almost complete. The epithelium extends over the damaged dermis and partially covers the developing granulation tissue. Blood flow into the area remains high.

(g) Eighteen days after injury. Subeschar sepsis has led to a gradual separation of the eschar. At this stage, granulating tissue becomes thicker and more vascular.

(h) Twenty-one days after injury. The sepsis has worsened and the wound is at its most vascular. Metabolic demands exerted by the wound are maximal.

(i) Twenty-eight days after injury. Complete separation of the eschar exposes the thick, vascular layer of granulating tissue which is heavily infected. Contraction of the myofibroblasts forces the collagen into whorls. The burn gradually becomes smaller, and the characteristic hypertrophic burn scar contracture develops.

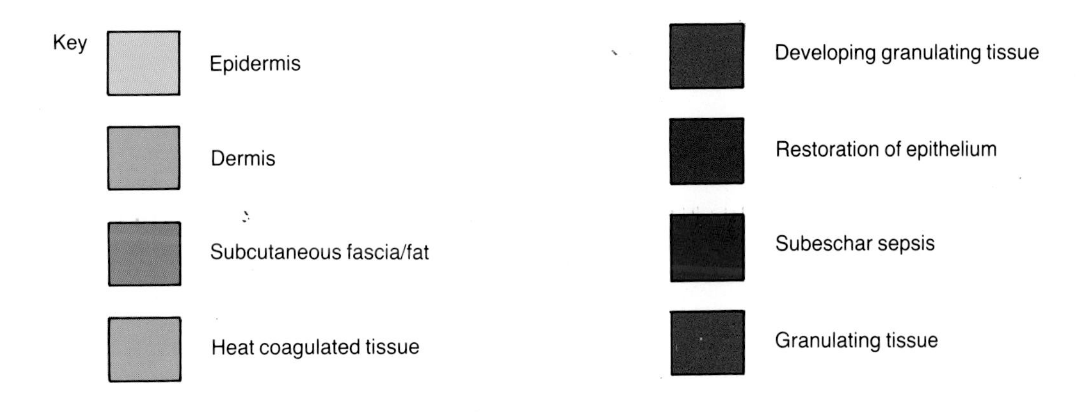

PARTIAL THICKNESS BURNS

Partial thickness burns can be subdivided into those involving the epidermis and superficial layers of the dermis (superficial dermal) and those reaching down into the deeper layers of the dermis (deep dermal). In both cases, spontaneous healing can occur from undamaged epidermal appendages – hair follicles, and sweat and sebaceous glands; epithelium will regenerate and resurface the remaining dermis (Fig. 1.2).

Superficial dermal

Superficial dermal burns blister, exposing the pink-coloured surface layers of the dermis which is sensitive to pain (Fig. 1.3). There is considerable loss of protein-rich tissue fluid from the wound, which when left to dry for two or three days will form an adherent scab. Spontaneous healing can occur within 10 to 14 days.

Fig. 1.2 Deep dermal burns after treatment in dressings for two weeks. White dermal net with a source of new epithelium from underlying hair follicles and sweat glands at the centre of each pink spot are seen.

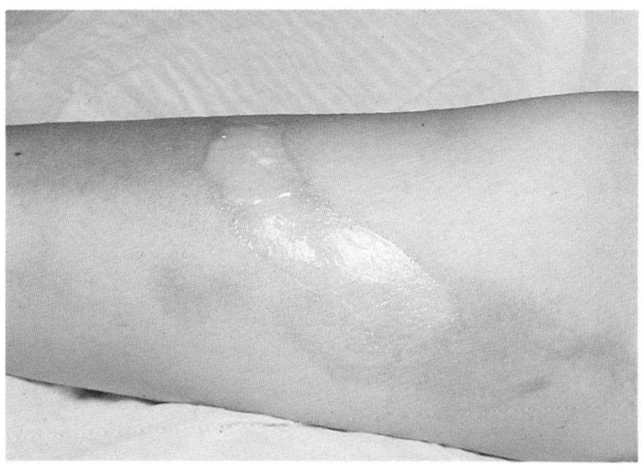

Fig. 1.3 Superficial dermal burns. (a) Very superficial burn. There is little erythema and only superficial blistering, as the burns were immediately cooled with running water.

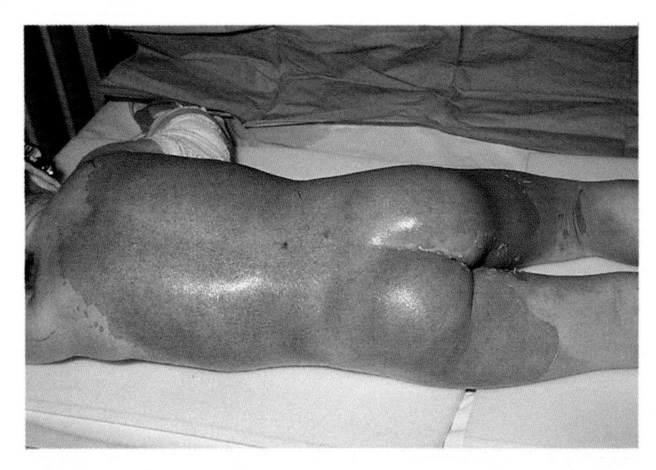

(b) The blistered skin has been removed to expose a moist, pink base which is very sensitive to touch.

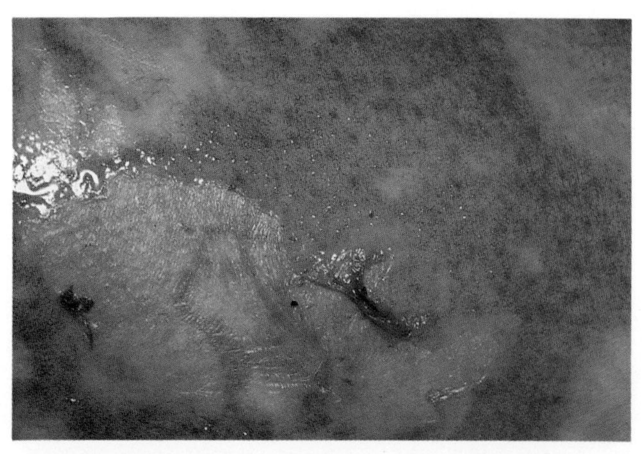

(c) The pink and red mottled dermal base has been exposed. Dessication following exposure to dry air promotes scab formation with some deepening of the dermal damage.

Deep dermal

Deep dermal burns also blister but the base is creamy–white and not sensitive to pain as nerve endings have been damaged (Fig. 1.4). Overt fluid loss is not as excessive and it is easy to confuse with a more superficial burn when the wound has been allowed to dry. Spontaneous healing may be delayed for three to four weeks and is frequently followed by the development of a hypertrophic scar.

FULL THICKNESS BURNS

Full thickness burns do not blister as there are no patent capillaries remaining in the bed and coagulation destroys the entire dermis together with all epidermal cells. The surface of the burn is initially grey–white and leathery (Fig. 1.5), and is completely painless. If allowed to dry a transparent brown eschar forms, revealing thrombosed vessels beneath (Fig. 1.6). With very deep burns, damage can extend to muscle and bone, thus adding a volume component to the injury (i.e. not only depth but also volume of dead tissue).

Only small, deep burns will heal spontaneously. The granulation tissue which eventually forms in the wound contracts (Fig. 1.7), making it smaller, allowing the surrounding undamaged epithelium to migrate onto the surface of the burn. In larger deep burns, healing is by secondary intention only and skin grafting is invariably required.

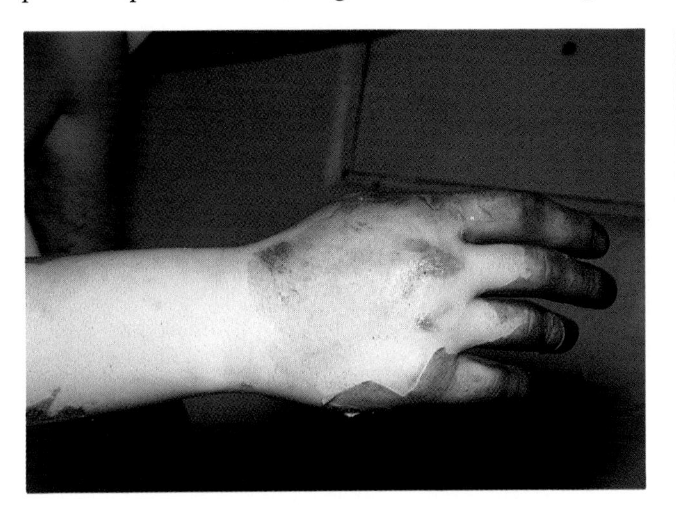

Fig. 1.4 Deep dermal burn. The dry white dermal bed is not painful to pinprick, yet not entirely anaesthetic. Spontaneous healing may be slow and followed by hypertrophic scarring.

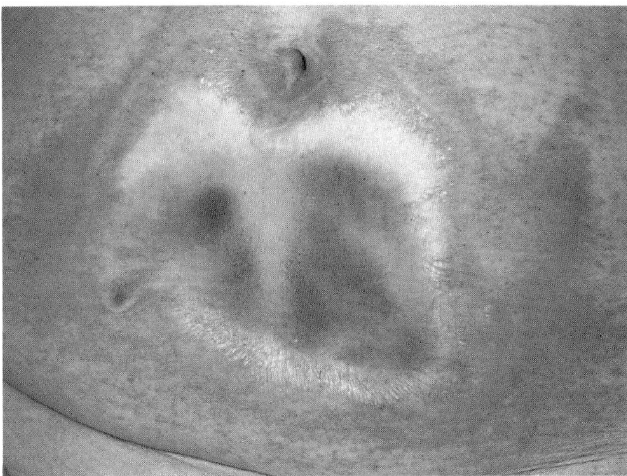

Fig. 1.5 Full thickness burn. There is a numb central leathery area, with white deep dermal damage around the periphery. The vascular flare is a zone of hyperaemia and indicates superficial damage.

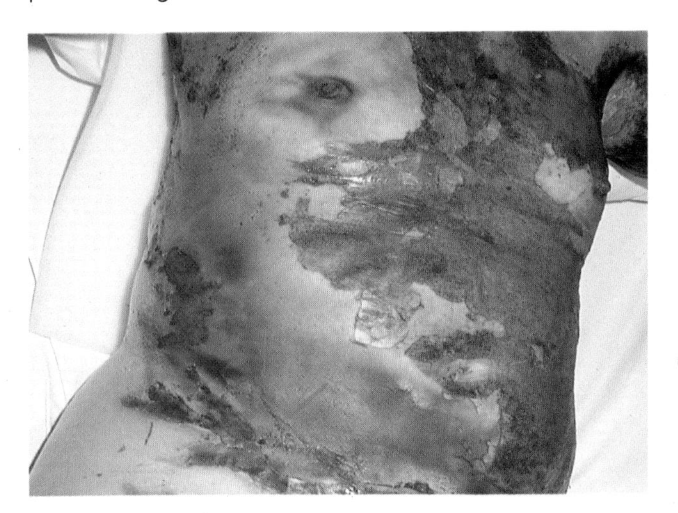

Fig. 1.6 Deep burn. Some areas are leathery and dull, and others parchment-like and transparent, revealing thrombosed vessels beneath.

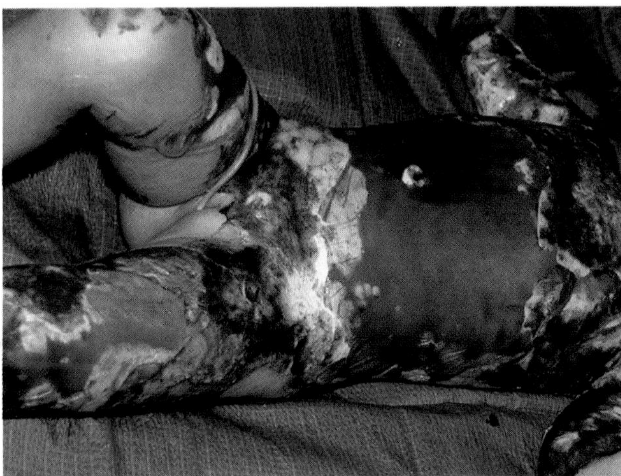

Fig. 1.7 Deep flame burn. The dry and leathery eschar around the abdomen is impairing respiration. Flexures show cracking and splitting.

BURN DEPTH

Assessment

The depth of a burn can be suggested by the circumstances of the accident, such as the causative agent and the duration of contact. Scalds, which are injuries caused by wet heat, tend to be superficial, although they will be deeper where contact is prolonged (Fig. 1.8). Flame burns are usually deep (Fig. 1.9), particularly where clothing has been ignited. Flash burns can appear severe, but although the temperature involved may be high, it is applied for only a fraction of a second with the deeper dermis invariably unaffected.

An enquiry into first aid measures taken immediately after the injury is essential. Rapid immersion in cold water cools tissues and limits or even reverses some of the cellular damage. It is necessary to establish the actual time of a burn injury, as the nature of the wound surface changes with time, thus altering the characteristic diagnostic appearances (Fig. 1.10).

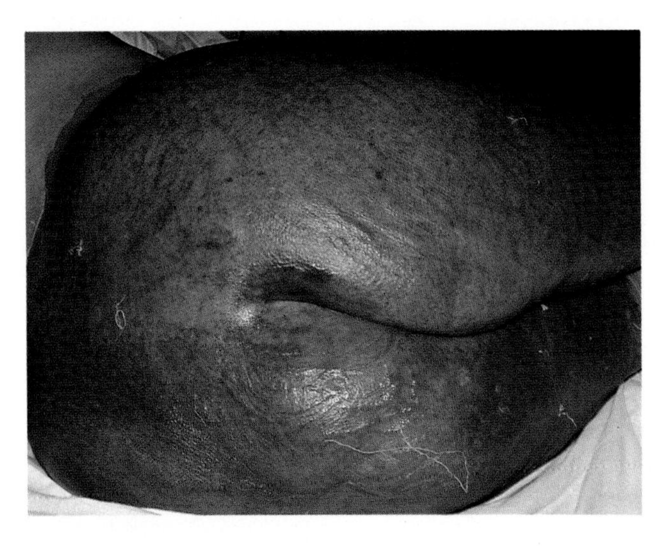

Fig. 1.8 Immersion scald. A deep dermal burn caused by prolonged contact with very hot water. The dry red base is stained by haemoglobin pigments.

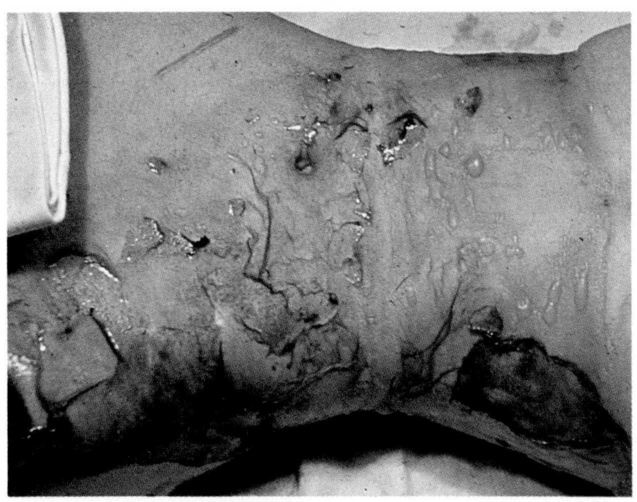

Fig. 1.9 Mixed depth flame burn. Areas of erythema, superficial and deep dermal burns, together with smoke staining of the skin, indicate full thickness skin loss.

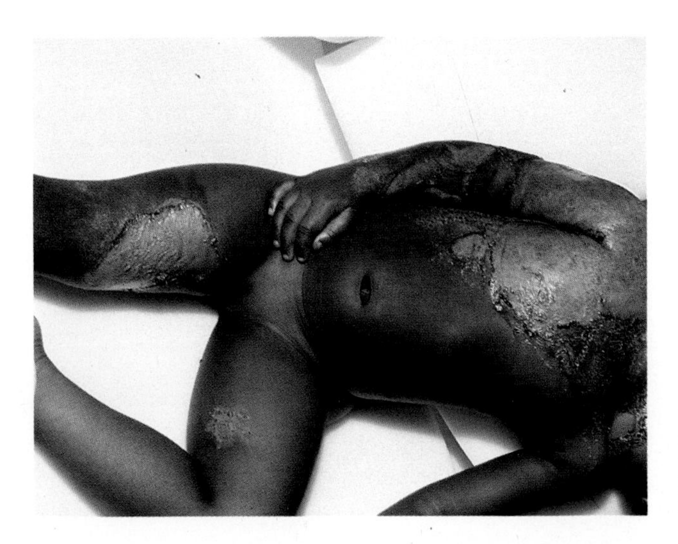

Fig. 1.10 Typical scald from hot water. The burn is 24 hours old, hence the surface is no longer moist and glistening. Because of delayed presentation, the level of injury is more difficult to assess.

Appearance

It is often remarkably difficult to judge the depth of a burn. Assessment may be made on the presence or absence of blister formation and also on the colour of the wound, although these are not necessarily definitive. Blistering implies that fluid is formed through damage to the capillaries. However, absence of blistering suggests that the underlying capillaries have been thrombosed and that the damage is more severe, involving the full thickness of the dermis. These factors are summarized in Table 1.1.

Pinprick test

Light pressure with a sterile needle will establish the degree of sensitivity. This test requires co-operation from the patient and is thus of limited value in unconscious patients or small children.

Inability to perceive pain indicates that nerve endings have been destroyed and that the burn is likely to be deep, whereas sharp appreciation of pain suggests that the wound is superficial. Dullness to pain but awareness of touch is the most reliable objective test of a deep dermal injury.

AREA

Prompt assessment of an area of burn is essential, as burns of an extent greater than 10 per cent in children and 15 per cent in adults warrant an immediate intravenous infusion. The Wallace Rule of Nine is convenient when estimating a burn area in adults (Fig. 1.11a), but it is less applicable in young children due to different proportions of head to trunk (i.e. larger head compared with trunk area). For children, the Lund and Browder chart is more relevant (Fig. 1.11b-,c). As a general guideline, the area of the patient's palm is equal to 1 per cent of body surface area.

As skin thickness varies in different parts of the body it is important to have a good working knowledge of these variations to assess a particular burn area. Mortality from a burn is related to the patient's age and the surface area involved, and a reasonably accurate estimation of the chances of survival can be obtained from the Bull and Fisher mortality tables (Fig. 1.12).

Table 1.1 Classification of burns

Type of burn	Blistering	Appearance	Pinprick test
Superficial dermal	Present	Bright red	Sensitive to pain
Deep dermal	Blisters are broken	Creamy coloured/ mottled	Dullness to pain, sensitive to touch
Full thickness	Absent	Grey/white or brown	No sensation

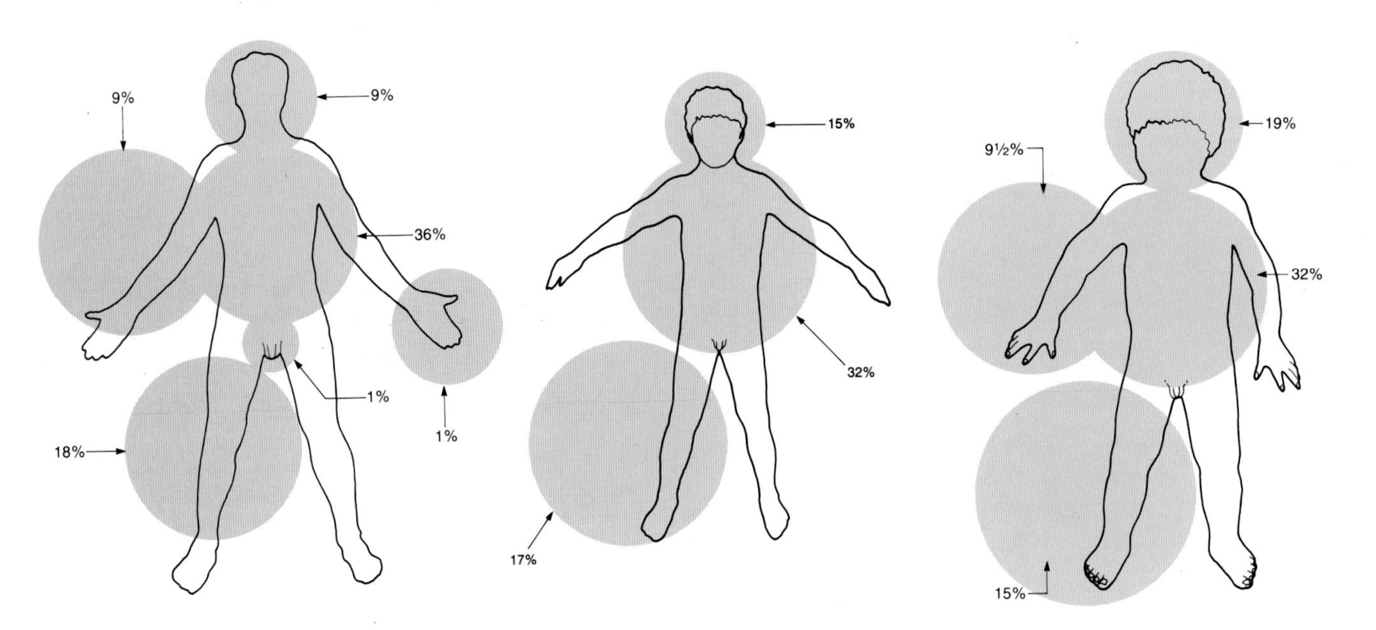

Fig. 1.11 Estimation of burn area. (a) Wallace rule of nine. (b) Lund and Browder modification of area in children of five to ten years of age. (c) Modification for children of up to four years of age.

Body area burned %	Age (year)																
	0–4	5–9	10–14	15–19	20–24	25–29	30–34	35–39	40–44	45–49	50–54	55–59	60–64	65–69	70–74	75–79	80 +
93 +	1	1	1	1	1	1	1	1	1	1	1	1	1	1	1	1	1
88–92	0.9	0.9	0.9	0.9	1	1	1	1	1	1	1	1	1	1	1	1	1
83–87	0.9	0.9	0.9	0.9	0.9	0.9	1	1	1	1	1	1	1	1	1	1	1
78–82	0.8	0.8	0.8	0.8	0.9	0.9	0.9	0.9	1	1	1	1	1	1	1	1	1
73–77	0.7	0.7	0.8	0.8	0.8	0.8	0.9	0.9	0.9	1	1	1	1	1	1	1	1
68–72	0.6	0.6	0.7	0.7	0.7	0.8	0.8	0.8	0.9	0.9	0.9	1	1	1	1	1	1
63–67	0.5	0.5	0.6	0.6	0.6	0.7	0.7	0.8	0.8	0.9	0.9	1	1	1	1	1	1
58–62	0.4	0.4	0.4	0.5	0.5	0.6	0.6	0.7	0.7	0.8	0.9	0.9	1	1	1	1	1
53–57	0.3	0.3	0.3	0.4	0.4	0.5	0.5	0.6	0.7	0.7	0.8	0.9	1	1	1	1	1
48–52	0.2	0.2	0.3	0.3	0.3	0.3	0.4	0.5	0.6	0.6	0.7	0.8	0.9	1	1	1	1
43–47	0.2	0.2	0.2	0.2	0.2	0.3	0.3	0.4	0.4	0.5	0.6	0.7	0.8	1	1	1	1
38–42	0.1	0.1	0.1	0.1	0.2	0.2	0.2	0.3	0.3	0.4	0.5	0.6	0.8	0.9	1	1	1
33–37	0.1	0.1	0.1	0.1	0.1	0.1	0.2	0.2	0.3	0.3	0.4	0.5	0.7	0.8	0.9	1	1
28–32	0	0	0	0	0.1	0.1	0.1	0.1	0.2	0.2	0.3	0.4	0.6	0.7	0.9	1	1
23–27	0	0	0	0	0	0	0.1	0.1	0.1	0.2	0.2	0.3	0.4	0.6	0.7	0.9	1
18–22	0	0	0	0	0	0	0	0.1	0.1	0.1	0.1	0.2	0.3	0.4	0.6	0.8	0.9
13–17	0	0	0	0	0	0	0	0	0	0.1	0.1	0.1	0.2	0.3	0.5	0.6	0.7
8–12	0	0	0	0	0	0	0	0	0	0	0.1	0.1	0.1	0.2	0.3	0.5	0.5
3–7	0	0	0	0	0	0	0	0	0	0	0	0	0.1	0.1	0.2	0.3	0.4
0–2	0	0	0	0	0	0	0	0	0	0	0	0	0	0.1	0.1	0.2	0.2

Fig. 1.12 Bull and Fisher mortality tables. This chart shows the relationship between mortality, age of the patient and total area of the body burned. 1 = 100% mortality; 0.1 = 10% mortality (from Bull, 1971).

2 Types of burn

SCALDS

Scalds are thermal injuries caused by wet heat, most usually boiling water or steam. Temperatures are relatively low and protein is denatured in the most superficial tissues (Fig. 2.1) although this is dependent on duration of contact (Fig. 2.2), first aid measures taken immediately after injury (Fig. 2.3) and the nature of the causative agent (Fig. 2.4).

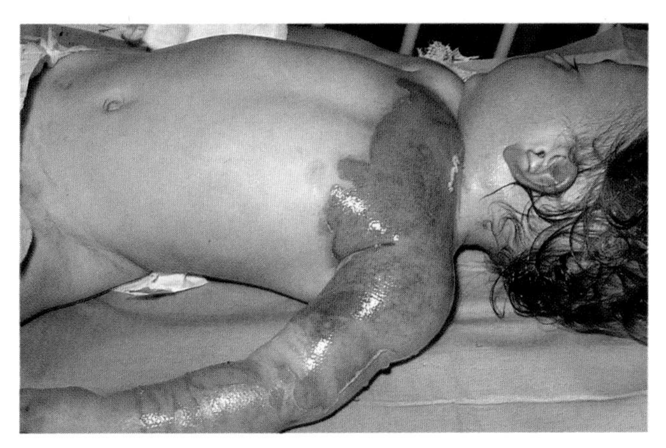

Fig. 2.1 Typical scalds from boiling water. (a) A characteristic distribution was caused on the face, neck, chest and arm from boiling water. The bright red base, which is moist and shining immediately after injury, suggests that the injury is superficial.

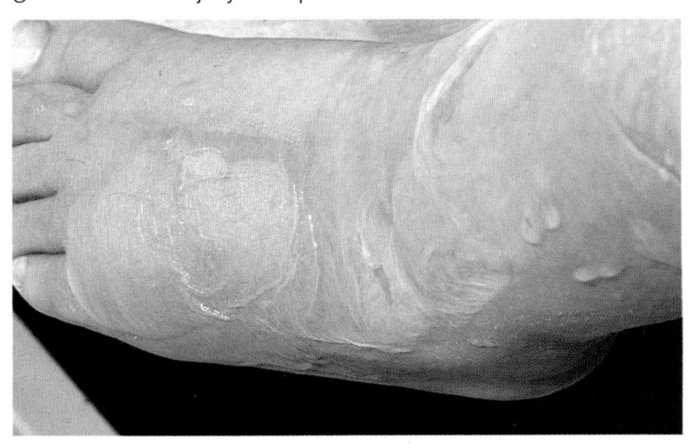

(b) A very superficial scald. This was seen very soon after the injury. There is little evidence of any serious exudate at this stage.

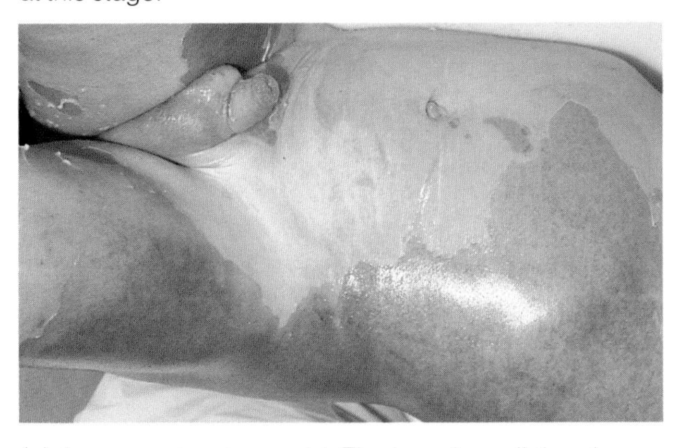

(c) A more extensive scald. The burn has all the characteristics of a superficial injury, but because of its extent intravenous resuscitation would be required.

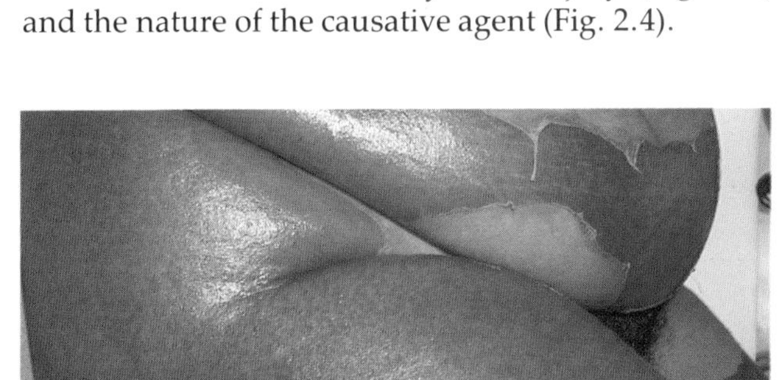

Fig. 2.2 Deep immersion scalds through prolonged contact with hot water. (a) The deep red base is anaesthetic and analgesic, suggestive of a deep burn.

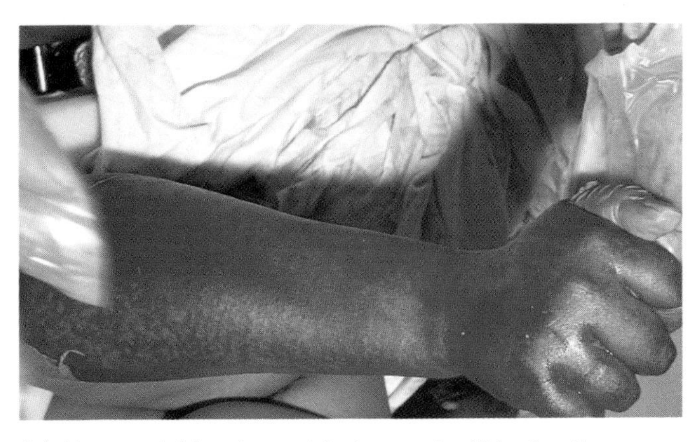

(b) Haemoglobin pigment is trapped within the tissues giving this typical dull, deep red appearance.

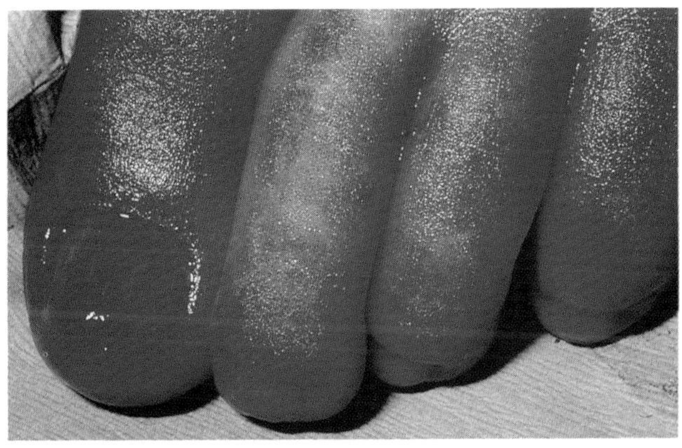

(c) The toenails have come away with the blisters, which usually indicates coagulation of the digital vessels and will result in loss of the toes.

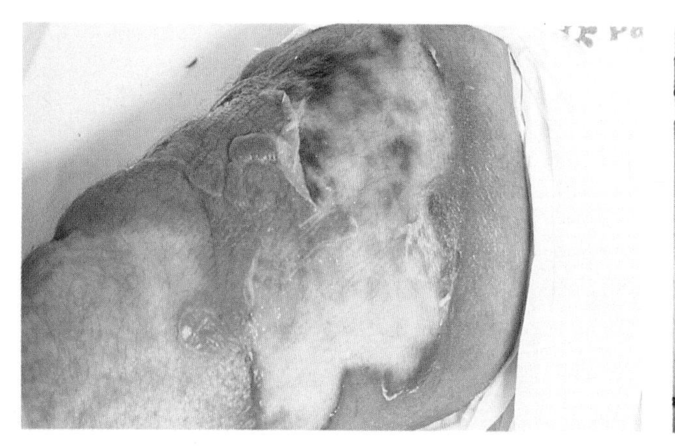

Fig. 2.3 Scald from boiling water. The white base is apparent immediately after injury and indicates a deep dermal or full thickness burn. Unfortunately, first aid measures of dousing with cold water were delayed and the burn needed skin grafting.

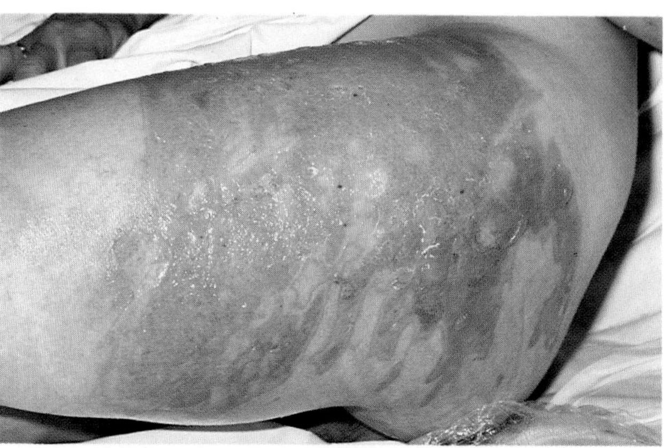

Fig. 2.4 Other agents causing scalding. (a) Boiling milk. The blistering is patchy and care should be taken to ensure that the unblistered areas are not deep, by testing sensation. Healing should be uneventful.

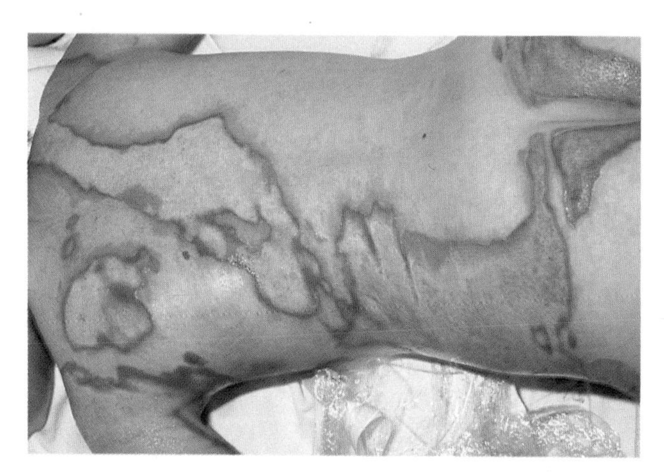

(b) Hot oil. The noticeable red margin with paler dermal base is characteristic of a deep dermal burn, and the distribution corresponds to the pattern of clothing. First aid measures were delayed.

(c) Hot fat. The base is pale and mottled, with patchy areas of numbness. The burns were dressed and healed spontaneously.

FLAME

These injuries result from the direct effect of burning gases on the skin. The temperatures involved are very high and coagulation of protein is greater than in scalds. Such burns are deep (Fig. 2.5), often with smoke staining of the skin (Fig. 2.6); a dry eschar develops through which thrombosed vessels are seen (Fig. 2.7). Prognosis is dependent on the extent of area burned (Fig. 2.8). Flame burns may often present with inhalational injuries (Fig. 2.9); early surgery may be contraindicated in the presence of significant pulmonary complications.

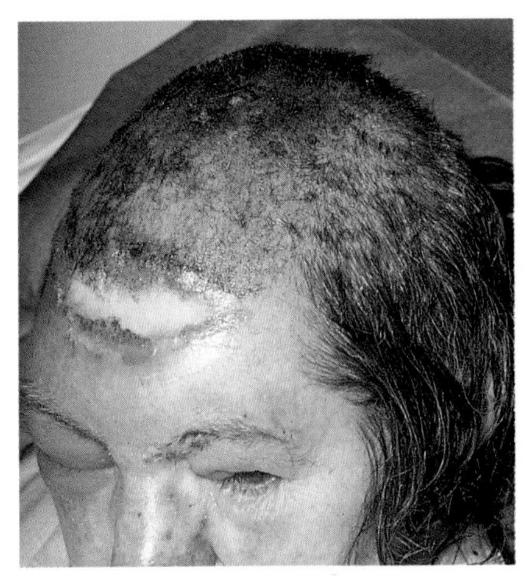

Fig. 2.5 Deep flame burns. (a) Burns caused by ignition of hair. Deep burns of forehead and scalp are seen with oedema of the eyes and nose. Hair should be shaved, but early surgery is often disappointing.

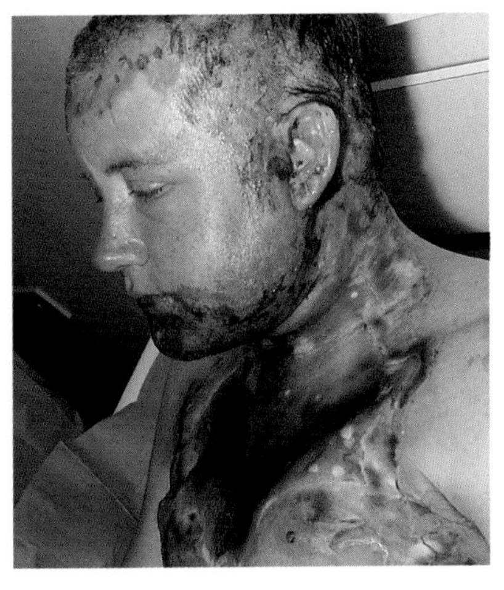

(b) Deep burn caused by ignition of alcohol. With such a burn over the neck early excision and grafting is important to protect access to airways.

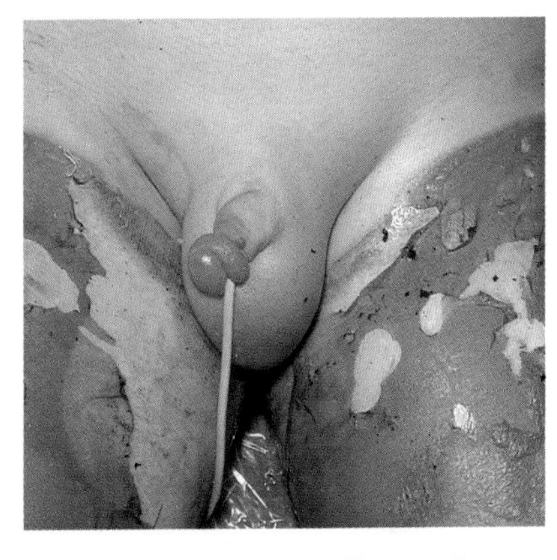

(c) Patchy blistering indicates that there is little capillary function remaining in the skin, and the anaesthetic white base is revealed in places. Local oedema is typical and the hourly output of urine is monitored. Micturition is not impaired by penile oedema, which is usually confined to the foreskin only.

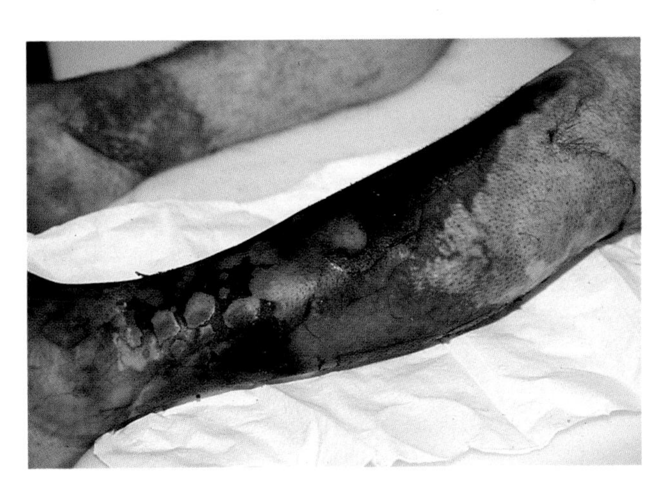

Fig. 2.6 Deep charring burns. Very deep circumferential burns of the leg. With this type of injury, underlying muscle damage may be anticipated, and immediate surgical decompression including fasciotomy is required.

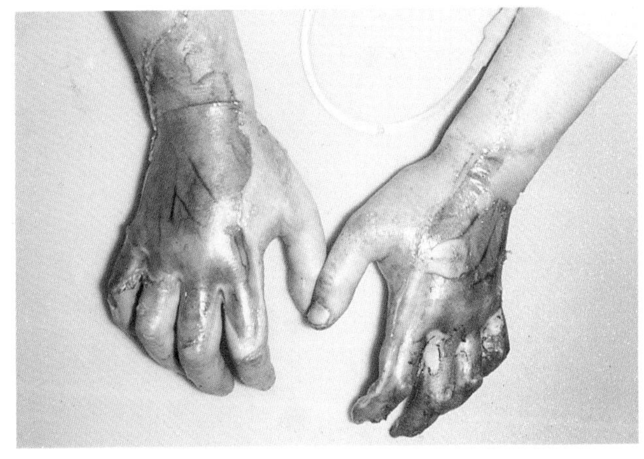

Fig. 2.7 Full thickness skin loss. The transparent eschar, 24 hours after admission, reveals thrombosed veins beneath, which is a typical characteristic. Sharp excision and skin grafting should be performed without delay.

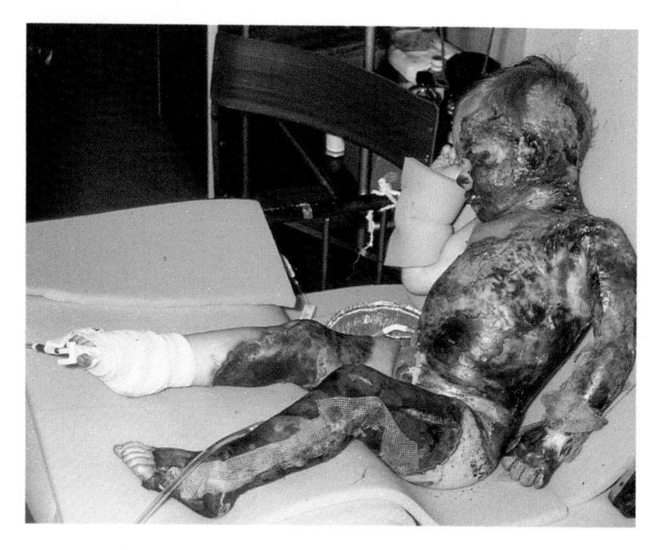

Fig. 2.8 Massive flame burn. Treatment is by exposure. An escharotomy on the leg is dressed with a strip of nitrofurazone, and the wound surface is clean and dry. Over 80 per cent of the body surface is deeply burned and prognosis is very poor indeed.

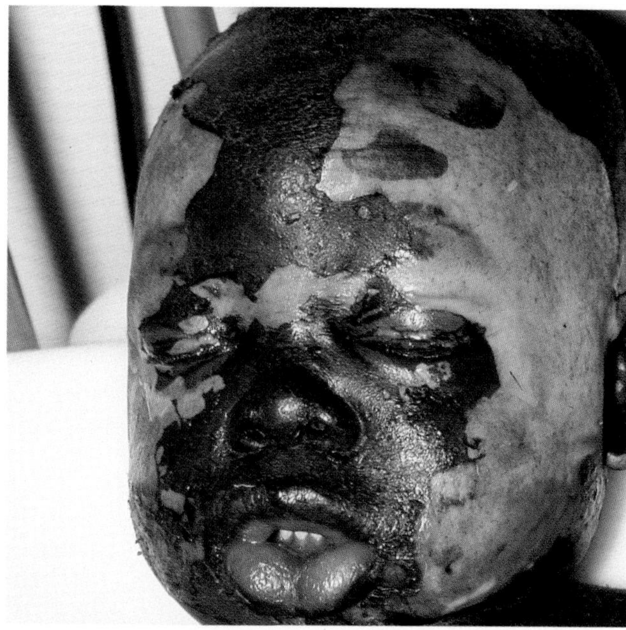

Fig. 2.9 Mixed depth burns with inhalational injuries. The cream coloured areas are deep and require grafting. Oedema of the lips, together with a stridor, indicate an upper airways obstruction, and early endotracheal intubation is prudent. In this case, injuries were caused by a gas explosion in a confined area.

CONTACT

Contact burns follow the direct conduction of heat from a hot surface to the body. This contact is usually short lived, with rapid reflex withdrawal. In circumstances where such protective mechanisms are not working, such as the patient suffering a blackout (Fig. 2.10), cases of alcoholism, epilepsy (Fig. 2.11) and drug addiction, or where the hand is mechanically trapped (Fig. 2.12), damage will extend much more deeply, often reaching bone (Fig. 2.13). Contact burns from molten bitumen, or tar often cause deep burns and physical removal of the agent may worsen injuries (Fig. 2.14). Solvents are themselves irritant and should be avoided. A frequently applied flamazine dressing allows the bitumen to separate spontaneously.

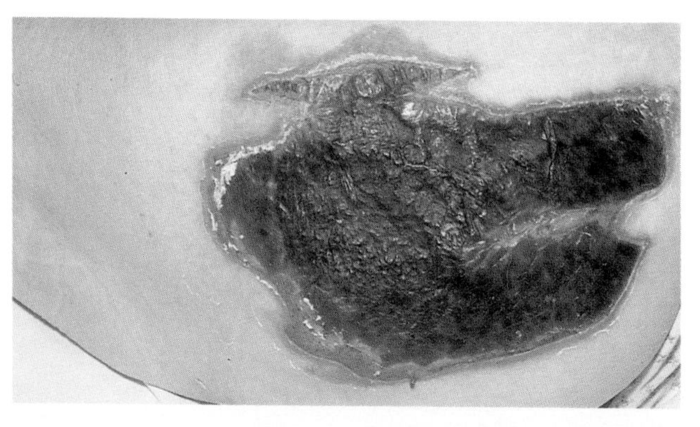

Fig. 2.10 Full thickness contact burns, sustained when protective mechanisms were not functioning. (a) Although water temperature in radiator pipes may not be very high, prolonged contact at temperatures above 45°C combined with the effects of local pressure cause full thickness burns of this type.

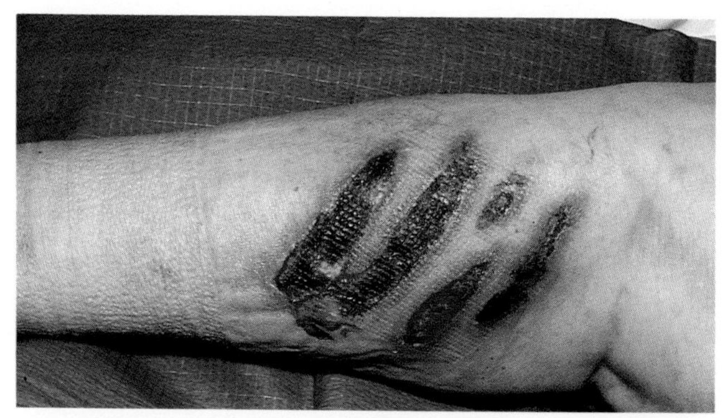

(b) Burns are deep due to prolonged contact, but their narrow extent allows conservative treatment.

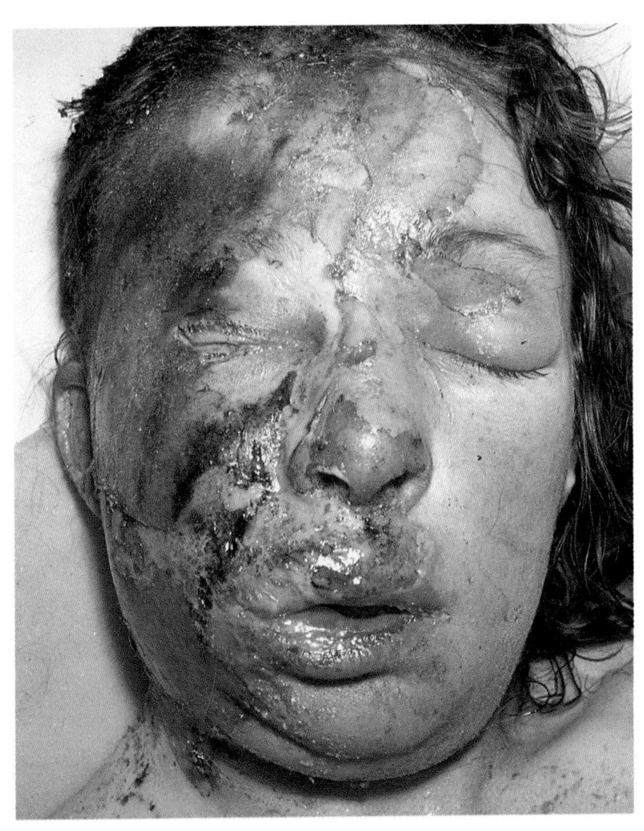

Fig. 2.11 Mixed flame burns with a deeper central area from prolonged contact with fire. The patient had an epileptic fit whilst ironing in front of an open fire.

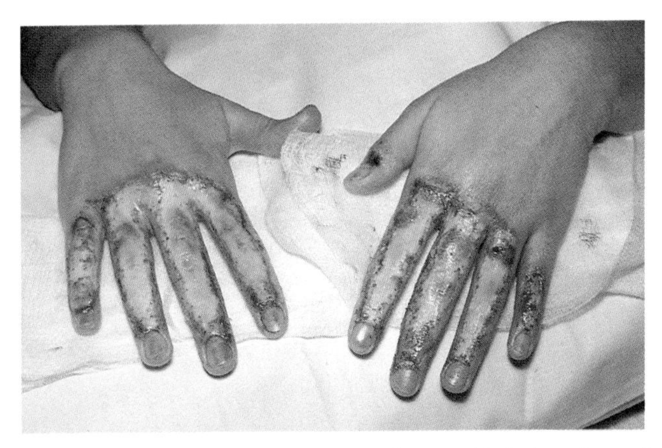

Fig. 2.12 Full thickness contact burns. Both hands had been momentarily trapped within a hot press. Although this burn is not particularly severe, prolonged contact at very high temperatures will cause devastating injuries.

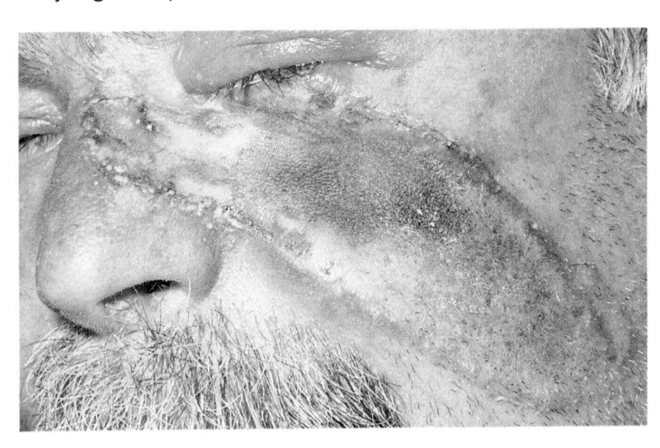

Fig. 2.13 Contact burn from a hot water pipe. These are typically deep burns which may extend down to zygoma, requiring flap cover.

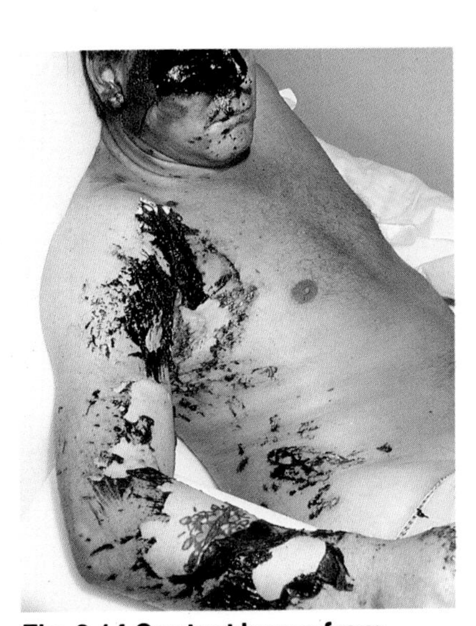

Fig. 2.14 Contact burns from molten bitumen, tar and metal. (a) Molten bitumen burns.

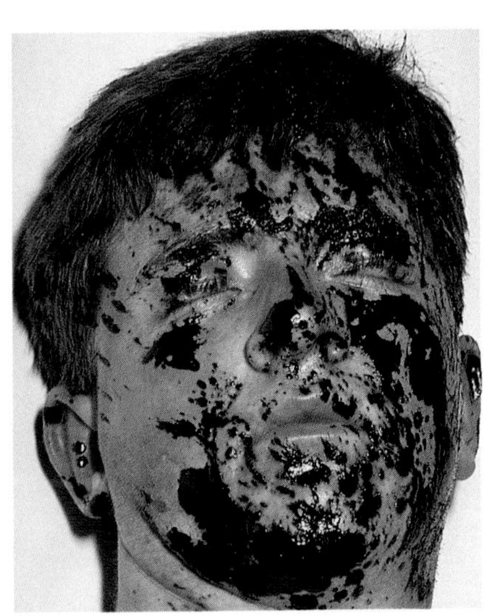

(b) Tar burns. Typical splashes on the face. Immediate first aid measures should be taken by thorough irrigation with water.

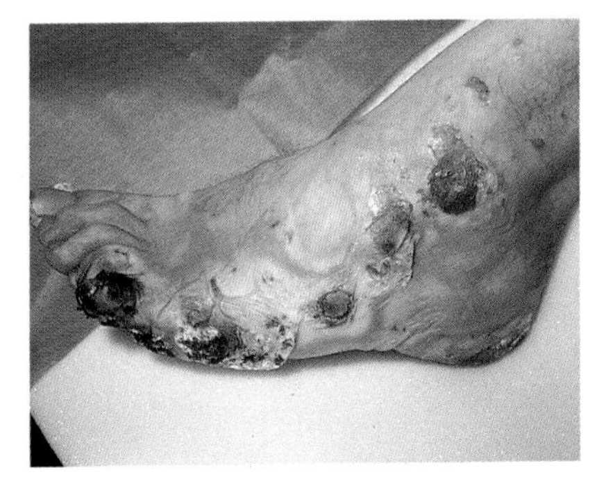

(c) Full thickness burn from molten metal. The hot metal melted the shoe in this instance. The injury will require skin grafting.

FRICTION

Friction burns often involve large areas of abrasion with some thermal damage caused by generated heat. Genuine friction burns are seen where the body is in contact with a rapidly moving mechanism such as a tyre or belt drive. These burns are deep and may expose fat, tendon or bone. Early primary excision may be necessary (Fig. 2.15).

ELECTRICAL BURNS

In electrical burns, the severity of damage depends upon the resistance of tissues (high resistance in bone and skin), the voltage and amperage (Fig. 2.16). This electrical energy is transformed to thermal energy, often with explosive force (Fig. 2.17).

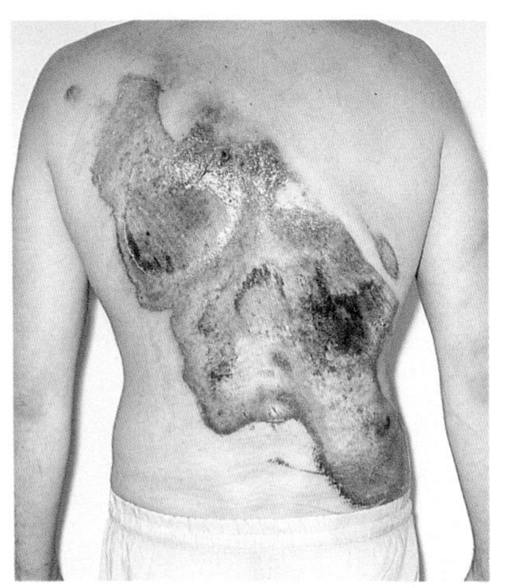

Fig. 2.15 Extensive friction burns. (a) Full thickness skin loss of the back, caused by the patient being dragged along a road for some distance.

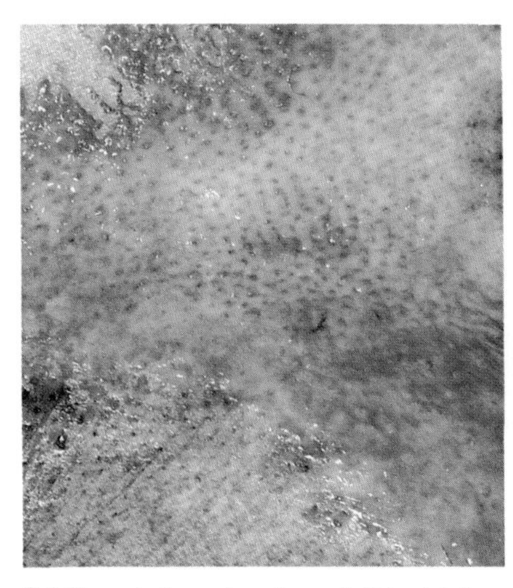

(b) Deeply ingrained road dirt which cannot be excised by shaving with a skin graft knife, as the grit quickly causes the blade to blunten. Removal is best performed with a wire dermabrasion brush.

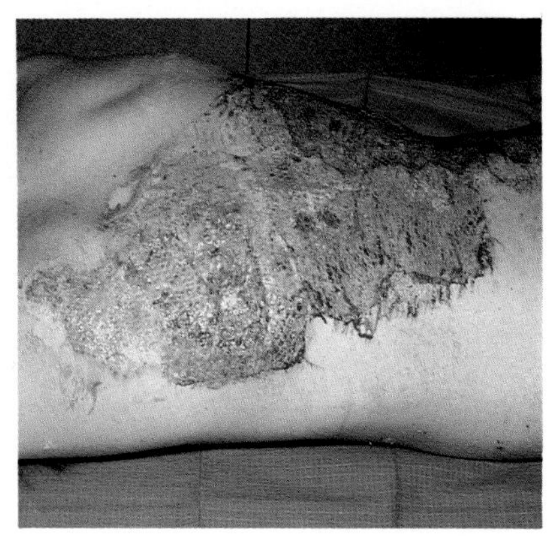

(c) Six days post-injury, following abrasion and thin skin grafting.

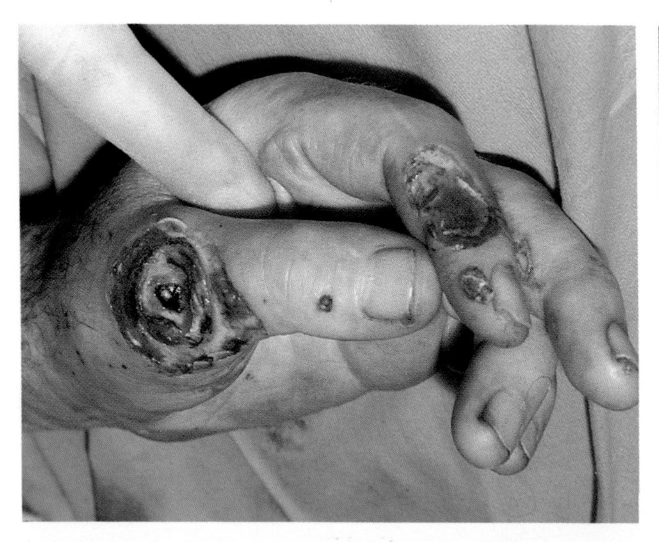

Fig. 2.16 Deep burns to the hands. Caused by contact with a cable carrying 660 volts. Damage is more extensive than with ordinary domestic voltage. The metacarpophalangeal joint will need cover with a local flap.

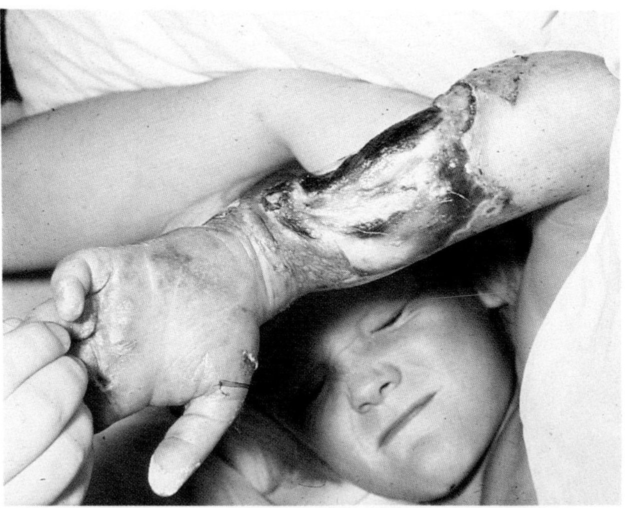

Fig. 2.17 Electrical injury with explosive force. Bare bone without periosteum is exposed but the hand, although insensate, still retains an adequate circulation. Late thrombosis can be anticipated and the arm in this case was amputated.

At ordinary domestic voltages the size of a cutaneous burn may appear small but is always deep, with a surprisingly large volume of tissue damaged (Fig. 2.18). Small punctate burns (Fig. 2.19) often mask thrombosed vessels and devascularized fat (Fig. 2.20). The body becomes part of an electrical circuit, and it is at the point of entry of the current that the wound is most severe (Fig. 2.21). Current passing across the heart can induce fatal ventricular fibrillation or cardiac arrhythmias. Similarly, flow across flexural surfaces can cause arcing; tetanic spasms may prevent the victim from releasing contact and may even be strong enough to rupture muscles (Fig. 2.22). Tetanic spasm of respiratory muscle may cause respiratory failure.

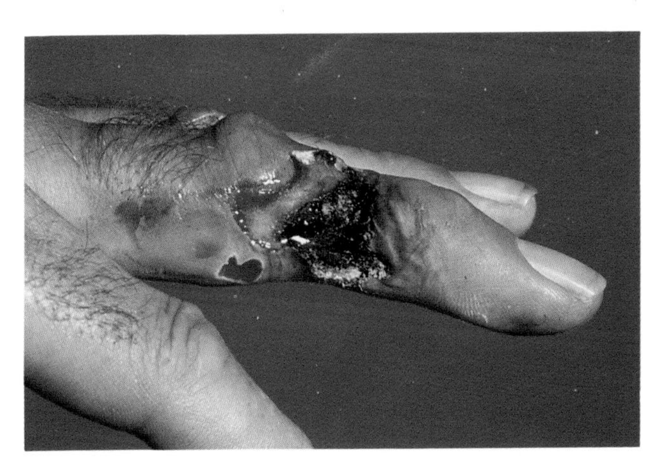

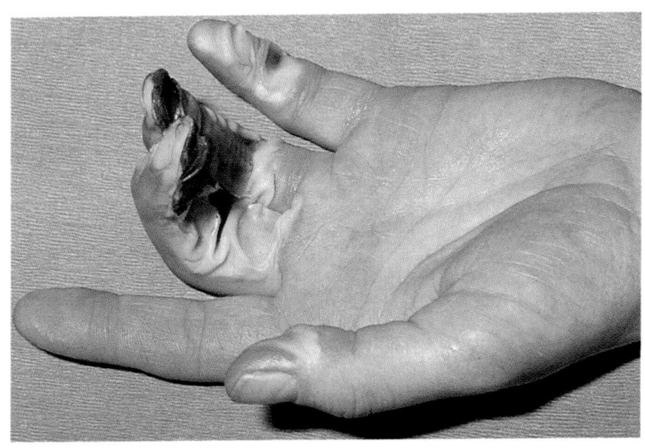

Fig. 2.18 Very deep electrical burns. (a) This injury involves nerves, blood vessels, tendons and bone and will require wide excision and flap cover. (b) The fingers have been mummified and dessicated. Both middle and ring fingers will require amputation.

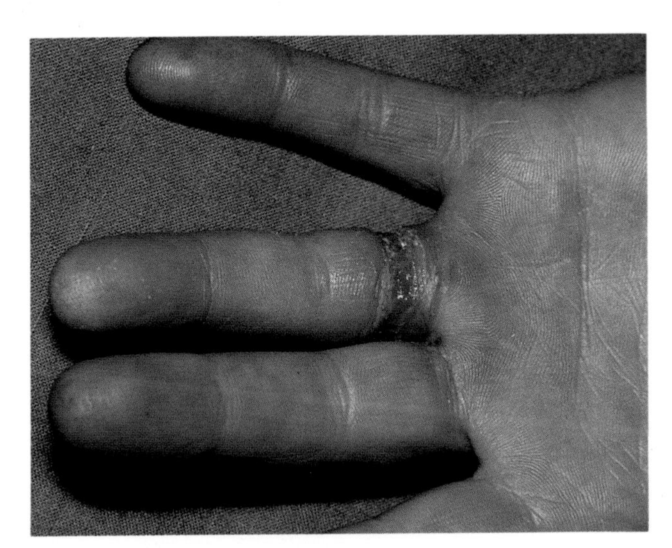

Fig. 2.19 Electrical contact burn. This injury, sustained through wearing a wedding ring, is fortunately minor; in the majority of cases the underlying neurovascular pedicle is thrombosed with eventual loss of the finger.

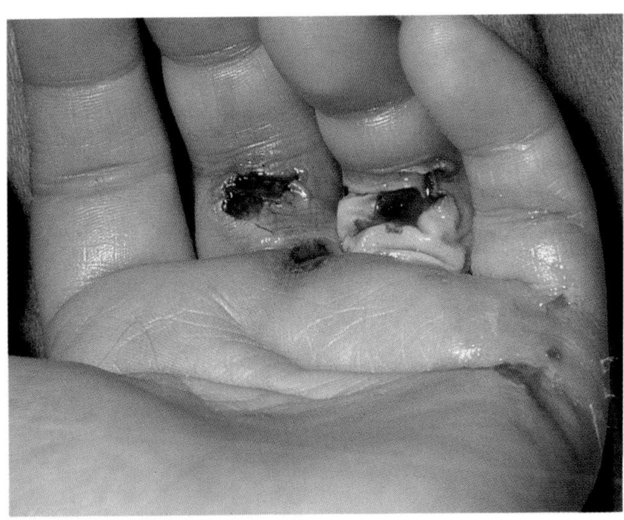

Fig. 2.20 Small punctate burns. (a) The peripheral blisters have been removed, revealing the full thickness central loss. These burns were treated with zinc oxide tape, allowing almost full return of function.

Injuries occurring in infants caused by chewing or sucking on live leads or plugs, where moist tissues decrease resistance to the passage of current, are always extensive (Fig. 2.23). Secondary haemorrhage from the labial artery can be dramatic.

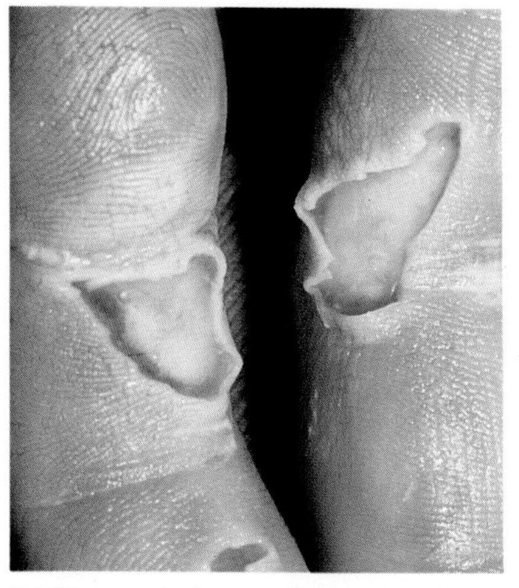

(b) The central area shows charring. These injuries may be treated conservatively.

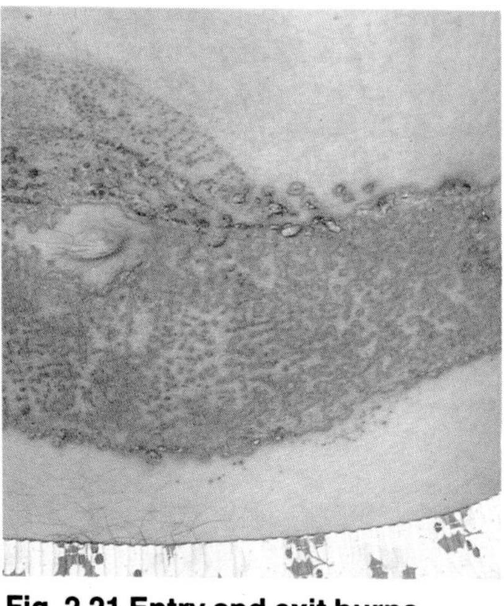

Fig. 2.21 Entry and exit burns.
(a) The exit burns on the abdomen were superficial and irregular.

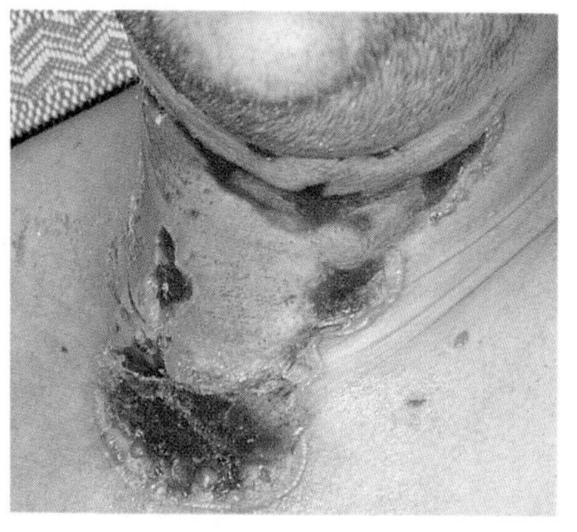

(b) The entry burn in the neck was deep, and a muscle flap was required to cover the exposed inominate vessels. These injuries were caused by a fall onto an electrified railway line.

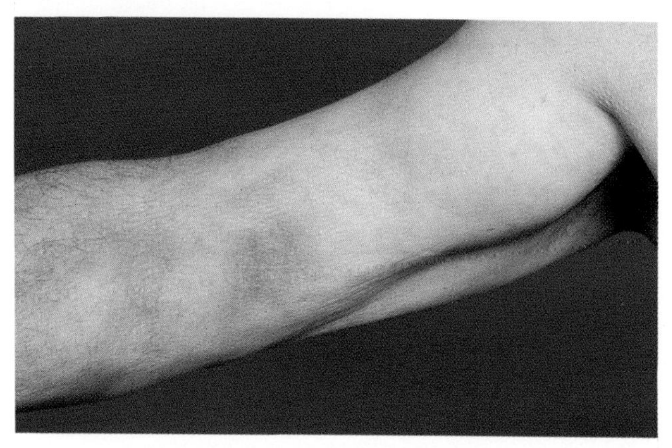

Fig. 2.22 Electrical injury. The biceps tendon has been ruptured following tetanic contracture. Note slight bruising of the antecubital fossa and contracted muscle belly in the upper part of the arm.

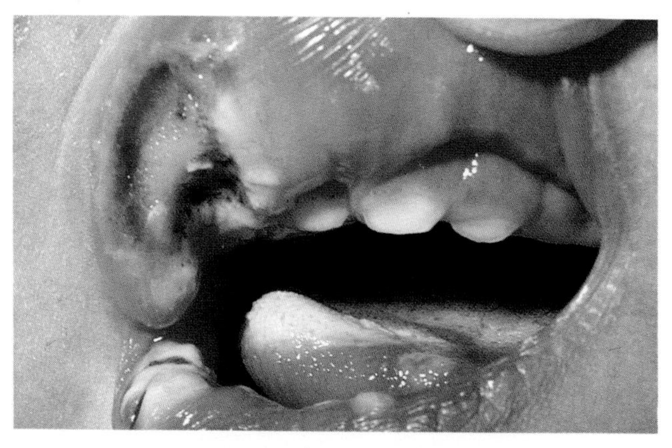

Fig. 2.23 Deep burns of lips and tongue. (a) Injuries were caused by sucking on a live plug.

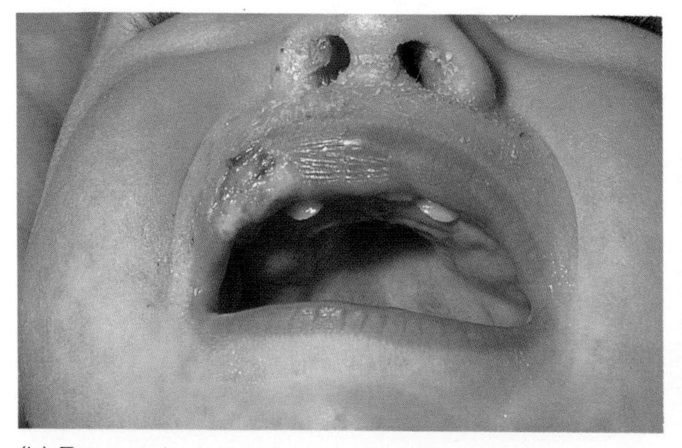

(b) Two weeks later, after conservative treatment. Healing is almost complete.

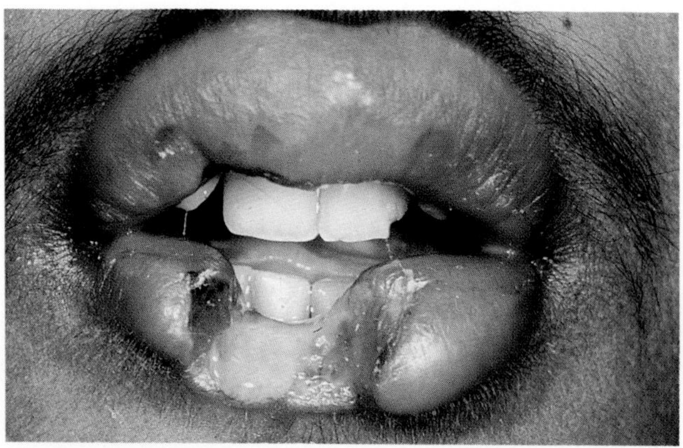

(c) This injury was caused by chewing an electric cable. There has been local arcing with extensive tissue loss. Secondary haemorrhage from the labial artery is a risk.

In the case of flash burns from arcing, where temperatures may exceed 4000°C for a fraction of a second, no current flows through the body but the impact may throw the patient several feet. Damage to the tissues is therefore superficial (Fig. 2.24) unless clothing has been ignited. However, damage to the eyes, when the blink reflex is too slow, may be more severe (Fig. 2.25), albeit very unusual.

High tension injuries of greater than 1000 volts result in massive muscle damage and charring, thrombosis of vessels and often fractures caused by the fall (Fig. 2.26). The volume of tissue damaged is always extensive (Fig. 2.27), and the temperature of these tissues may take hours to return to normal. Surgical decompression with excision of coagulated muscle, as well as wide fasciotomies and escharotomies should be undertaken urgently and the wound inspected repeatedly. Late thrombosis occurs with the apparent extension of the area devitalized, and secondary haemorrhage is common.

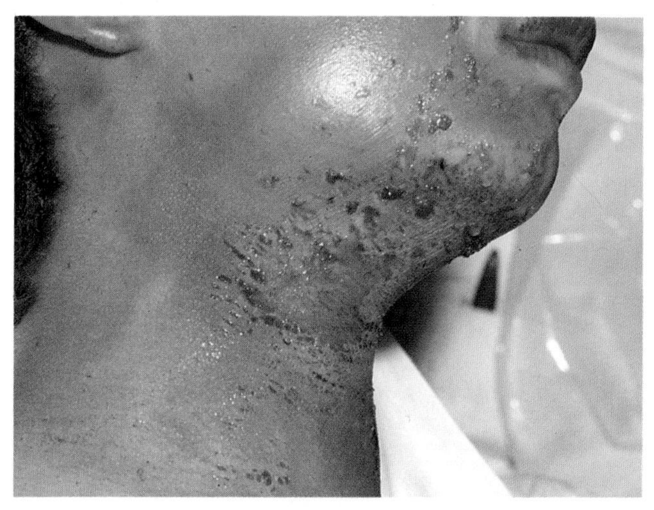

Fig. 2.24 Burn from high energy arc. Tiny punctate areas due to vaporized metal, sustained when the patient threw a metal chain across electrified rails.

Fig. 2.25 High energy flash burn. There is spotting of the skin from vaporized metal. Early conjunctival oedema and an apparently serious burning to the cornea which will, however heal rapidly. The blink reflex is sometimes too slow to protect the cornea in electrical flash injuries.

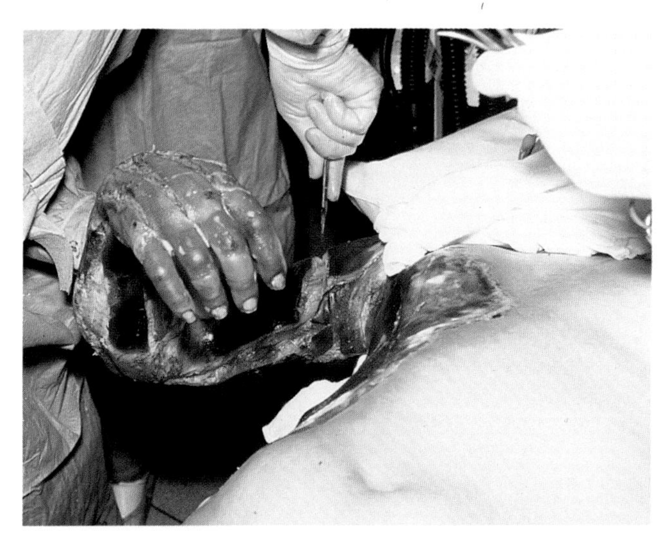

Fig. 2.26 A high tension electrical injury. (a) Circulation in the arm has been interrupted by thrombosis of the branchial vessels.

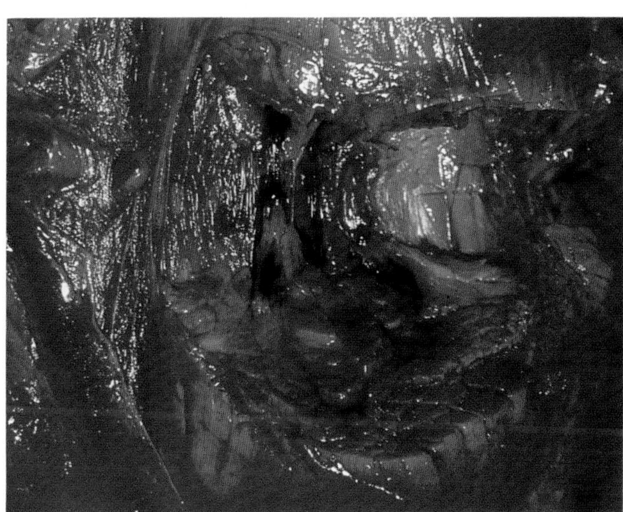

(b) An incision made through the biceps muscle at the time of amputation shows complete thrombosis of the vessels and coagulation of the muscle.

In thermoelectric burns tetanic spasms prolong contact with the source. In the common case of electric bar fires, massive damage to the palmar surface of the hand results in destruction of skin, tendons, nerves, vessels and even bone (Fig. 2.28). Skin grafting and flap cover is often necessary.

In lightning injuries, the current from a lightning strike can flow through the body but more commonly it flows across its surface, stripping off clothing. In these instances a fern-like pattern can be seen in the skin, which fades within hours. Cutaneous burns are usually small and neurological signs can be transitory. Paralytic ileus is seen, and cataracts may develop many years after the accident. Instant cardiac arrest is the usual cause of death although many cases of complete recovery have been recorded where cardiac massage and mouth to mouth resuscitation was persisted with for 30 minutes.

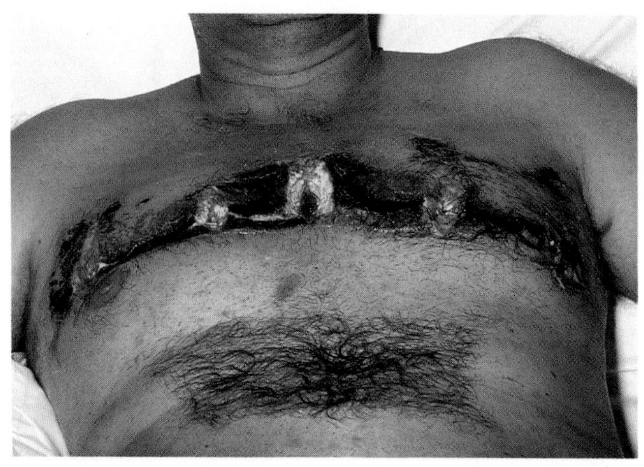

Fig. 2.27 Very deep electrical burn. This was caused by a fall onto a railway line. There is a very deep burn over the sternum, with loss of pectoralis major muscle and part of the sternum, thrombosis of the internal mammary vessels, and deep involvement of the pericardium. An omental flap was used to cover the exposed heart.

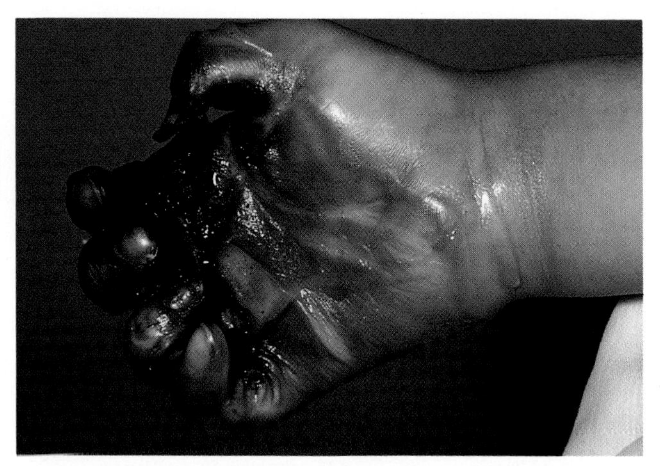

Fig. 2.28 Thermoelectric burns. The hand has been held in prolonged contact with a heated electrical element by a tetanic spasm. The tissue damage is always very extensive; all structures are involved and growth of the bones is invariably arrested.

CHEMICAL BURNS

Acid

The extent of injuries depends on the concentration of the acid, duration of contact and first aid measures taken. Acids cause a coagulative necrosis of tissues and with few exceptions (i.e. hydrofluoric acid; Fig. 2.29) do not penetrate deeply (Fig. 2.30). The acids are consumed and neutralized by the proteins themselves, and the burns tend to be deep and leathery, with a characteristic appearance (Fig. 2.31). First aid is by thorough irrigation with running water, although in acid burns to the eye blepharospasm, swelling and oedema may make adequate irrigation difficult (Fig. 2.32).

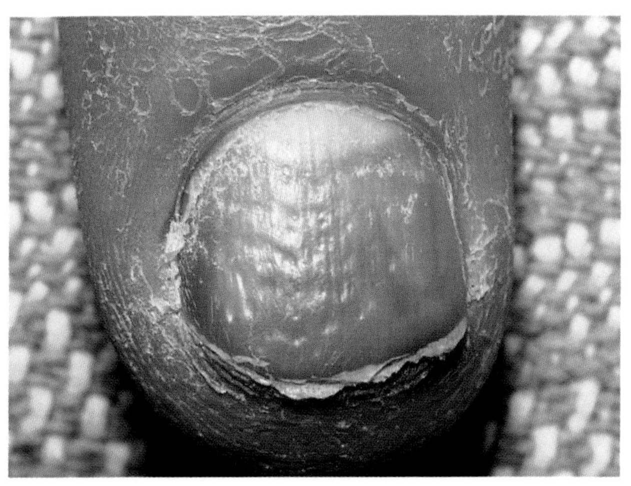

Fig. 2.29 Hydrofluoric acid burns. These are very corrosive and penetrating. Extensive irrigation should be carried out as soon as possible after contact. Calcium gluconate gel is widely used in limiting symptoms, but intradermal injection of calcium gluconate can cause extensive tissue necrosis.

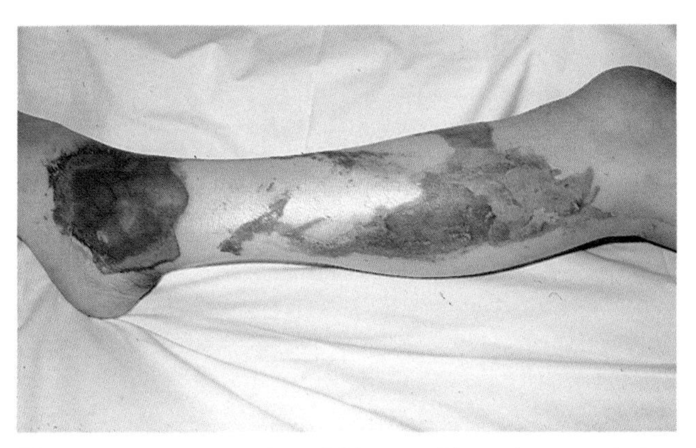

Fig. 2.30 Hydrochloric acid burns. There is coagulative necrosis with little penetration of deeper tissues. These burns are usually deep but do not worsen with time. The grey appearance is typical.

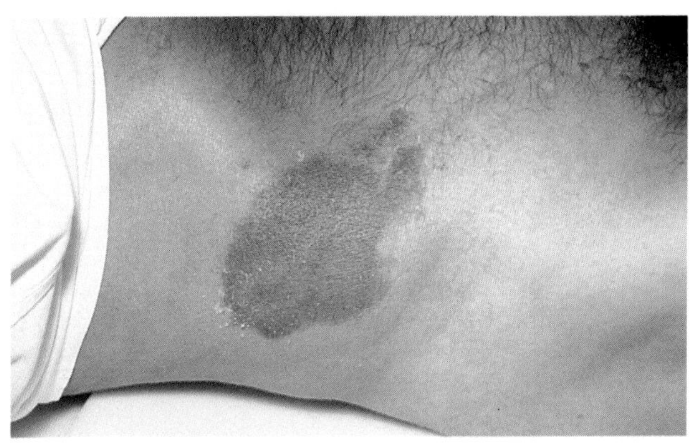

Fig. 2.31 Typical acid burns. (a) Sulphuric acid burn. Brown discoloration is typical. The coagulative necrosis limits penetration. This type of injury should be treated by irrigation with water.

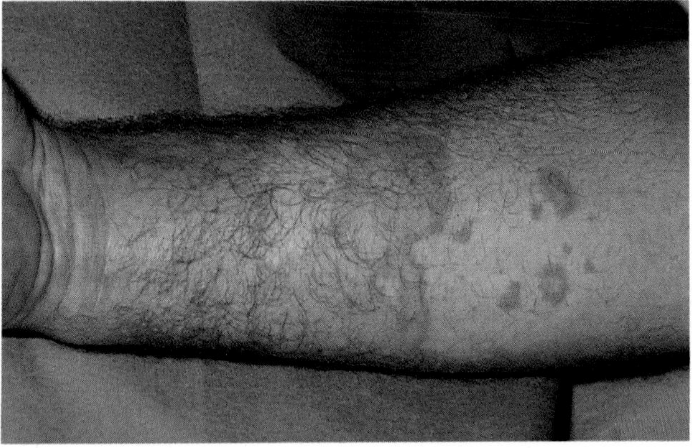

(b) Nitric acid burn. This was irrigated copiously for an hour and healed spontaneously, albeit with subsequent hypertrophic scarring.

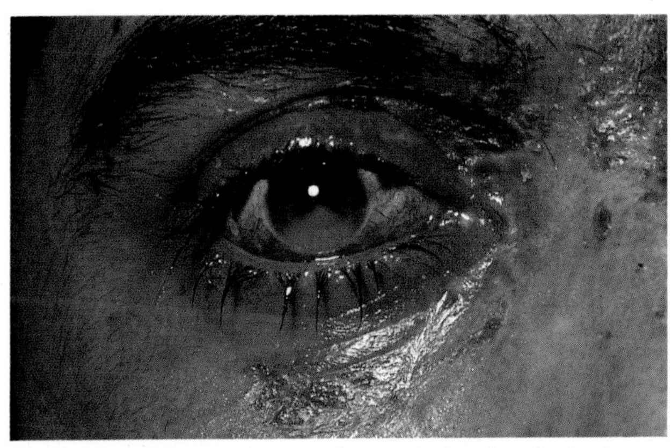

Fig. 2.32 Acid burn to eye. Corneal clouding in the lower half with some conjunctival irritation. Intense blepharospasm may prevent adequate irrigation.

Alkali

Alkali burns are usually more severe as they penetrate more deeply (Fig. 2.33). The colliquative necrosis does not isolate the chemical reaction and the alkali continues to penetrate (Fig. 2.34). First aid and initial therapy involve thorough irrigation, which should be prolonged for at least several hours. Attempts at chemical neutralization are also to be avoided as the exothermic reaction adds a thermal element to the injury. Cement burns are seen frequently; they usually present late and are common around knees and ankles (Fig. 2.35).

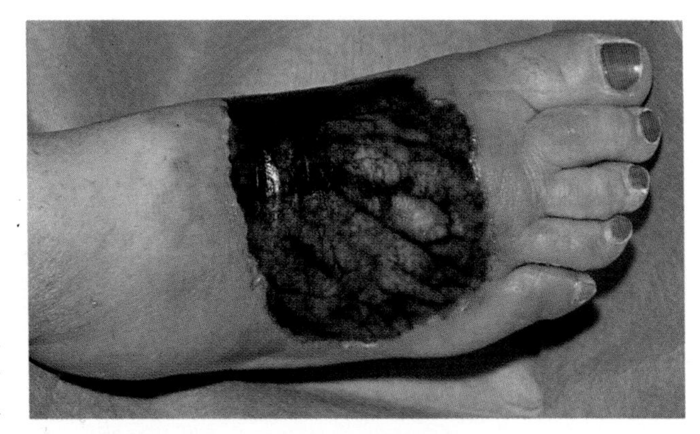

Fig. 2.33 Deep burn from oven cleaner. One week after the accident. There is obvious full thickness loss, and instant excision and grafting are indicated.

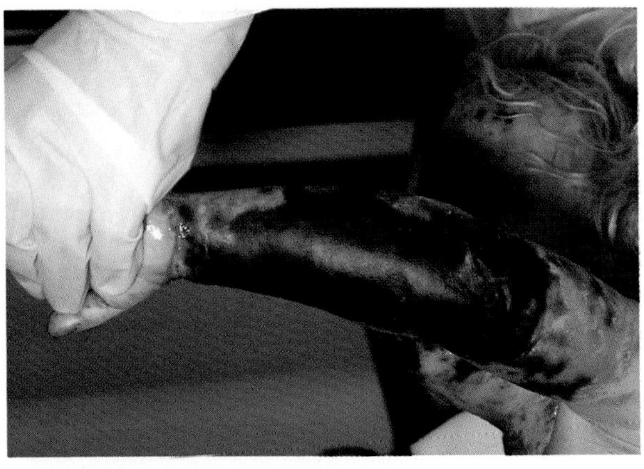

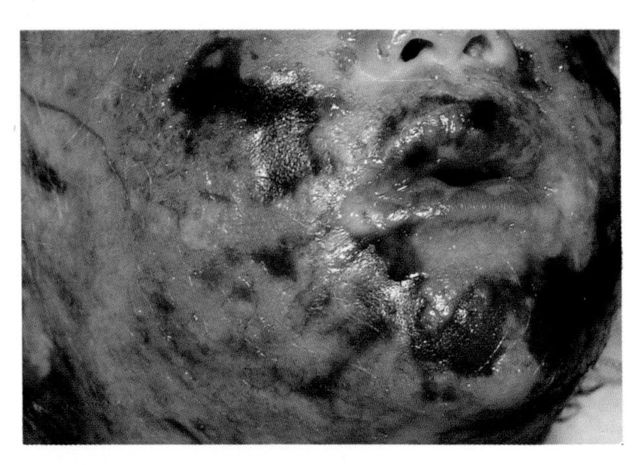

Fig. 2.34 Caustic soda burns.
(a) These are very deep and penetrating. In this case, first aid measures were delayed. Early excision and grafting are warranted.

(b) The black and brown areas are deep and will heal slowly, leaving bad scars.

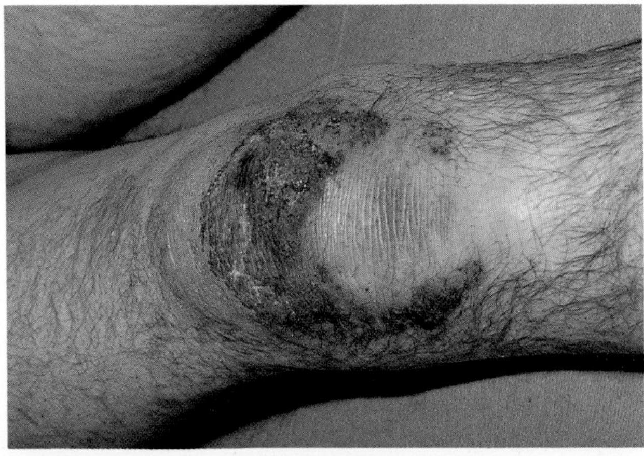

Fig. 2.35 Cement burns. (a) At the time of injury there is little pain experienced thus patients often present late. Management is by early shaving with delayed split skin grafting.

(b) This burn has been dressed with a simple greasy dressing for two weeks. The deeper layers of the dermis are exposed and there is obvious full thickness loss centrally.

Other chemicals

Many different chemicals can cause tissue damage but the principles of treatment are the same: thorough irrigation (Fig. 2.36), physical removal of the causative agent (Fig. 2.37), and prevention of infection with early excision and grafting if the lesions are insensitive to pinprick. Systemic absorption or inhalation of the chemical can cause serious metabolic or toxic problems (Fig. 2.38) and advice must be obtained from a toxicologist as early as possible after presentation.

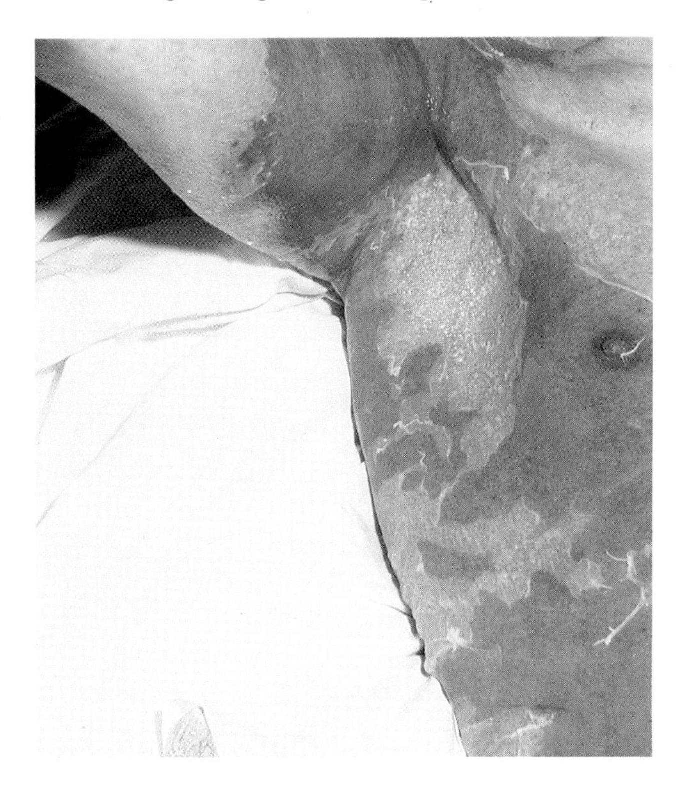

Fig. 2.36 Chemical burn from prolonged contact with petrol. These are usually superficial provided thorough irrigation and washing with soap and water are prompt.

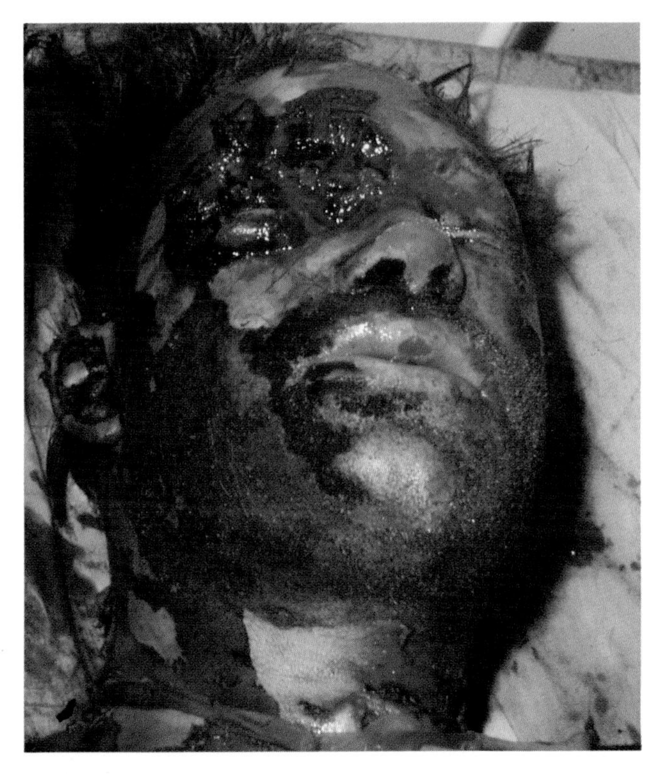

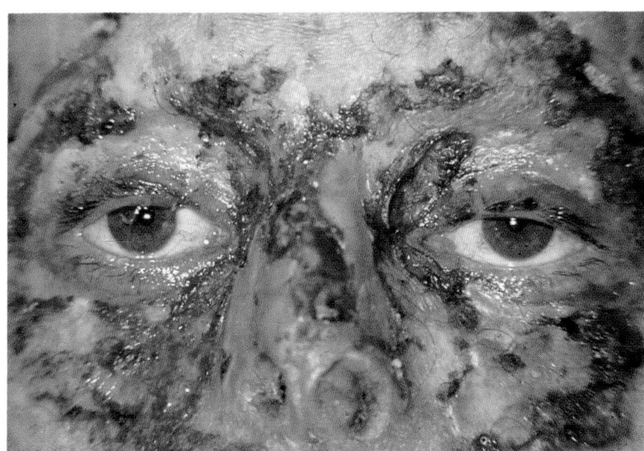

Fig. 2.37 Phosphorus burns. (a) Weak copper sulphate solution has been used to stain the particles black, to enable physical removal. Phosphorus continues to burn if exposed to air, therefore burns should be kept moist or wet dressings applied before the phosphorus particles are removed.
(b) Appearance some weeks later.

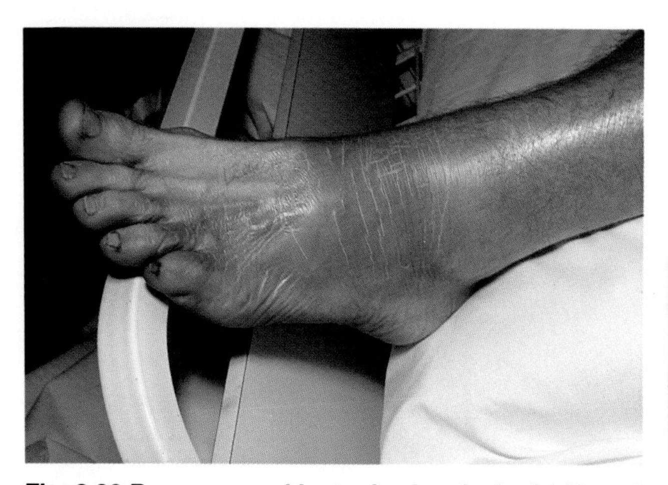

Fig. 2.38 Burns caused by toxic chemicals. (a) Phenol burns. Although locally toxic to the skin, phenol rarely causes full thickness injuries. This patient suffered renal failure and needed haemodialysis for six weeks to restore renal function.

(b) Mustard gas burns. Vesiculation with erythema and scattered deep dermal loss. Most of these burns healed spontaneously, albeit slowly. Grafting is rarely required, but changes in pigmentation are frequent. The inhalational component is much more important clinically.

3 Early management

Pathophysiology

Heat can cause complete destruction and necrosis of the dermis. Protein is coagulated and vasoactive substances are liberated into the local tissues, increasing the permeability of capillaries. This results in rapid loss of fluid and protein elements from the circulating blood into the interstitial tissues.

Consequently the burn wound becomes swollen and oedematous. The volume of fluid and protein loss from the blood becomes maximal at about eight hours after the incident. Fluid loss will continue for about 48 hours, by which time the capillary permeability has reverted to normal and extravasated water returns to circulation. Swelling is reduced and the wet, oozing surface becomes dry and leathery; an eschar will develop if no dressing is applied.

Over the next three to four days, there is a gradual increase in local circulation. With the ingrowth of capillaries into the boundary between dead and damaged tissues, the process of repair begins. At the same time, the undamaged peripheral tissues start to regenerate, with an increase in the rate of mitosis in epidermal cells and fibroblasts of the dermis.

The process of repair, with the formation of granulating tissue, fibrosis and scar, is concurrent with the process of regeneration and reconstruction of damaged tissues. Repair and regeneration can be affected by bacterial infection. Non-virulent or contaminant organisms delay the ability of the body to regenerate tissues. Virulent organisms damage the capability for regeneration and encourage the pathological processes of repair. The two processes of regeneration and repair are thus competing, i.e. the ability to heal with little destruction of the tissues, or the development of scars and contractures.

In the superficial wound, epithelial regeneration proceeds provided there is no contamination by virulent organisms. In the deeper wound, where the dermis is damaged, the process of repair with the formation of granulating tissue begins, and if the wound is not covered quickly by regenerating epithelium, considerable scarring will result. In the deepest wound, where all epidermal elements have been destroyed, healing can only be accomplished by granulation and wound contracture. Myofibroblasts in the granulating tissue rapidly decrease the area of the wound, thereby facilitating the slow spread of epithelium from the wound edges.

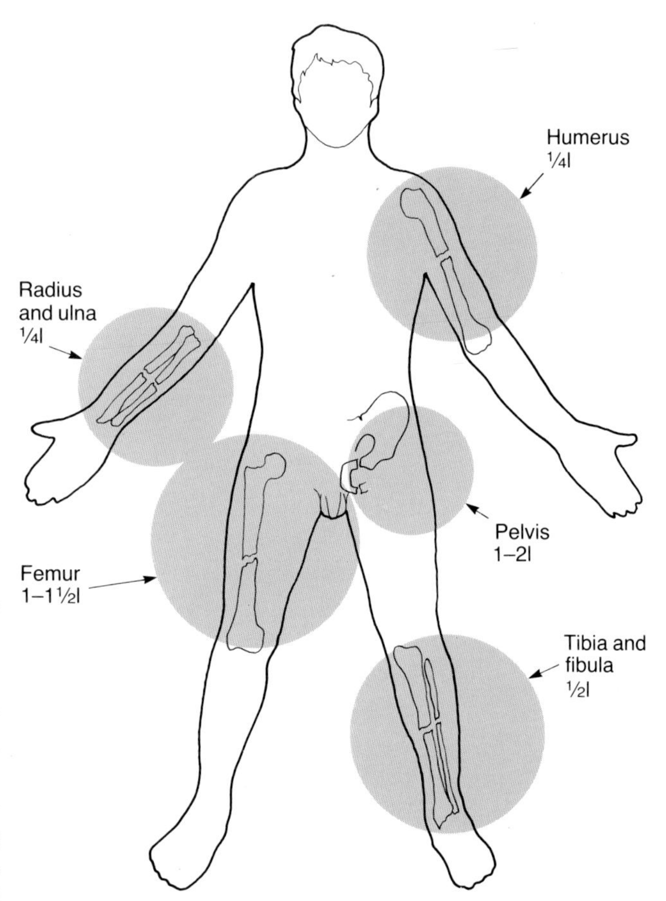

Fig. 3.1 Anticipated blood loss associated with fractures in different areas of the body.

RESUSCITATION

Initial fluid loss

Patients with burns involving more than 10 per cent of body surface area in children, or 15 per cent in adults, are likely to need intravenous fluid therapy. Those with smaller burns should only require extra water and sodium, although in infants care must be taken to avoid giving large amounts of electrolyte-free water as hyponatraemia and water intoxication can result. Preparations used in the treatment of infantile diarrhoea, e.g. Dioralyte and Moyer's solution (4g NaCl + 1.5g $NaCO_3$ per litre of water) are very effective. Generally, the amount of fluid lost is related to the area burned, but in cases where the area is greater than 30 per cent there can be a generalized increase in capillary permeability, with protein-rich fluid sequestered in unburned tissues, the mesentery of the bowel and the lungs.

Reduction in the circulating blood volume is due to loss of sodium, water and proteins; the rate of loss is greatest during the first eight hours after the burn injury, when capillary permeability is most marked. After approximately 36 hours, as capillary permeability returns to normal, fluid is drawn back into the circulation and the need for intravenous fluid support ceases.

In cases where fractures accompany burn injuries, pure blood loss from fractures as well as plasma loss from the burn must be taken into consideration. External bleeding prior to hospital admission should also be established. A guide to the blood loss incurred with fractures is shown in Fig. 3.1. An assessment of the amount of plasma and blood to be transfused should be reviewed frequently.

There is little agreement on the best method of resuscitating burn patients. Many methods are used and most are successful, but two prevail: the early use of colloids and the use of crystalloids. Both appear to be equally effective but patients treated with colloid make greater progress, and by being less oedematous have fewer problems with paralytic ileus, stress ulcers and pulmonary complications. However, colloids are expensive and plasma is not widely available although it is the most logical replacement fluid. Dextran is a suitable alternative.

Formulae give the physician only a guide to the average patient requirements (Table 3.1). Administered volumes should be adjusted according to the patient's response. Skin colour and warmth are good indices of peripheral circulation, and core and peripheral temperatures should be noted, together with the pulse rate and blood pressure. Increasing restlessness indicates that hypoxia is developing due to either an inhalational injury or, more often, simple hypovolaemia. Invasive monitoring with central venous lines is not recommended in uncomplicated burns as this can easily lead to the overprovision of fluids, but four-hourly estimations of haemoglobin and haematocrit are very useful. Measurement of the haematocrit, whereby a sample of blood is centrifuged down to enable the packed cell volume to be calculated, provides an indication of the efficiency of treatment (Fig. 3.2). A urinary catheter should be inserted for those patients requiring intravenous fluids, and hourly urine volume is the most informative guide to the adequacy of resuscitation. A fall in urinary output suggests that resuscitation is inadequate and should be challenged with rapid administration of 500ml of colloid. Diuretics should be avoided until it is considered that the patient has been overtransfused; most patients failing to respond to subsequent diuretics have not been given enough intravenous fluids. Early renal failure is seen only in patients where resuscitation has been delayed, where there is extensive muscle damage, or with pre-existing renal pathology.

WOUND CARE

After adequate sedation of the patient the wound should be carefully inspected in a clean and warm atmosphere. To minimize contamination, gloves, mask and apron should be worn. Any loose skin is gently removed and large blisters emptied; the area is cleaned with a mild detergent antiseptic and the surrounding skin shaved.

A burn can be treated by one of three methods: exposure to air; dressing; or prompt excision and grafting. Management options range from early radi-

Table 3.1 Formulae for adults, used in resuscitation

Crystalloids	Up to 24 h	Ringer's lactate (4ml/Kg/% Burn)	Half the volume in 8 h, remaining in 16 h
	24-48 h	5% dextrose 0.5ml colloid/% BSAB/kg	
Colloids	Up to 48 h	Dextran 110 in normal saline (120 ml/% BSAB = total volume*)	Half the volume in 8 h, quarter in next 16 h, remaining in next 24 h
Plasma	Up to 36 h, divided into 6 periods	$\frac{\text{\% BSA burns x wt(kg)}}{2}$ = ml required for each period	Period 1 = 4 h 2 = 4 h 3 = 4 h 4 = 6 h 5 = 6 h 6 = 12 h

*total volume should not exceed 6l

cal excision to benign neglect where dead tissue separates spontaneously. The condition of the patient and local facilities available will determine the course of treatment followed.

The 'exposure' method

This is based on the principle that drying the burn inhibits bacterial growth. Exposure is more appropriate for partial thickness or mixed depth burns where a decision has been taken to avoid early surgery. It is a difficult method to practise in circumferential burns as their surface will not remain dry. Favourable results have been reported by washing the wound once or twice a day with normal saline, but still maintaining a dry environment.

After exposing the wound in a warm, dry atmosphere an eschar should develop within two or three days, beneath which the wound will start to heal (Fig. 3.3).

The eschar forms a close-fitting cover consisting of protein exudate and dead, dessicated burned tissue which is lifted off as the burn heals. The dry eschar separates early, with good recovéry of the epidermal adnexae. The wound may become slightly deeper from dessication, but this is not of clinical importance. It is safe to allow spontaneous healing to proceed for two weeks, when any remaining eschar should be removed under general anaesthetic and raw areas grafted (Fig. 3.4). Eschar which remains in place for longer than this period will encourage bacterial colonization within the burn and lead to infection. If the wound becomes moist, fails to dry within two to three days or shows signs of infection (Fig. 3.5), the technique must be abandoned (Fig. 3.6) and

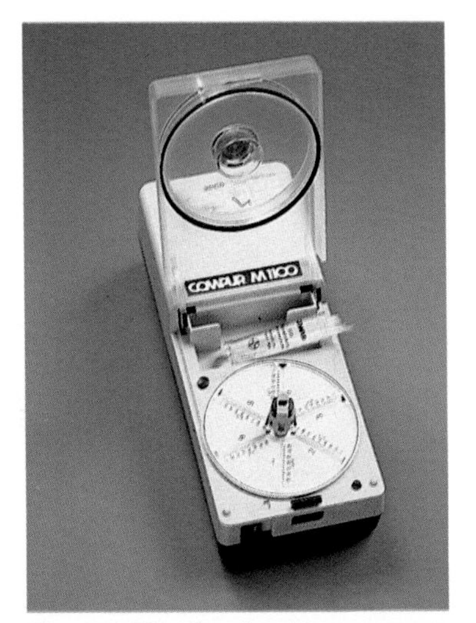

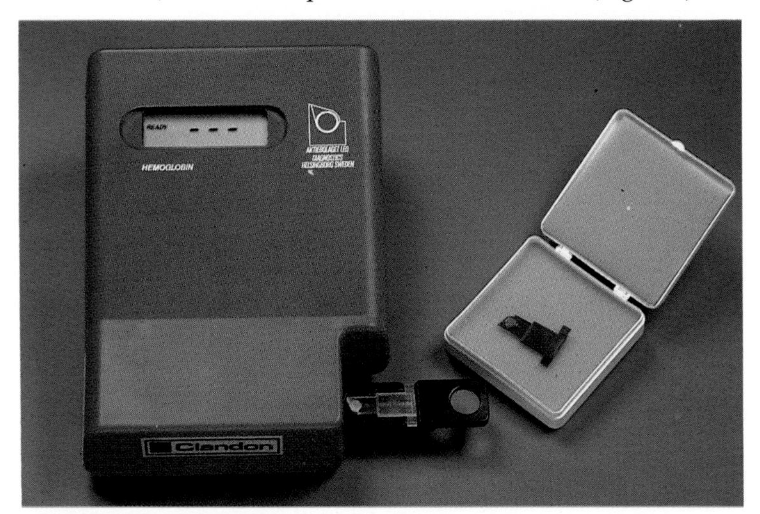

Fig. 3.2 Monitoring instruments used at the bedside. (a) Mini-centrifuge for measurement of the haematocrit. Blood, obtained by pricking the finger, is drawn into a capillary tube and spun down. (b) Haemóglobinometer.

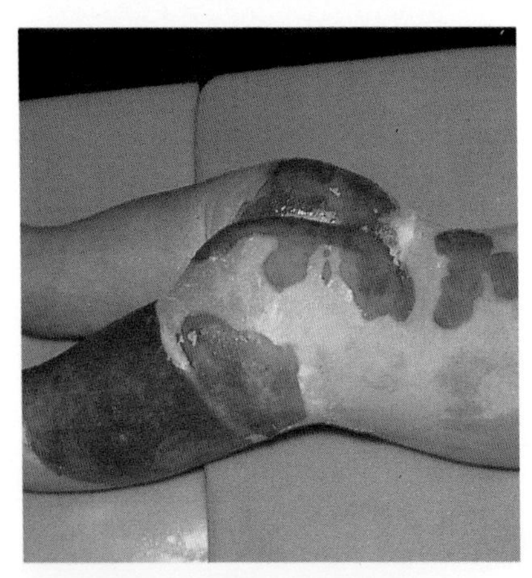

Fig. 3.3 Scald injury treated by exposure. (a) At 24 hours.

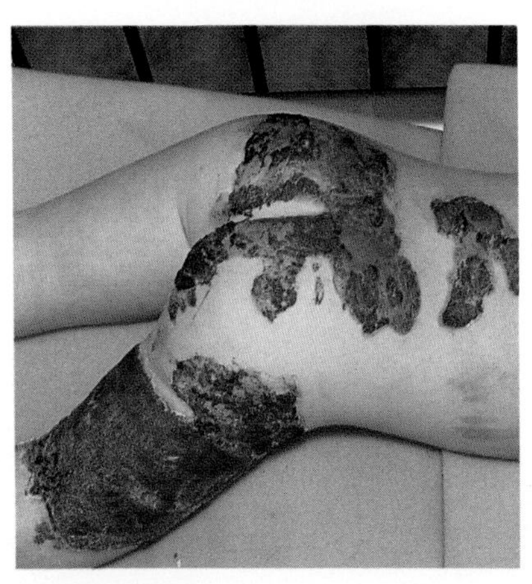

(b) At 7 days.

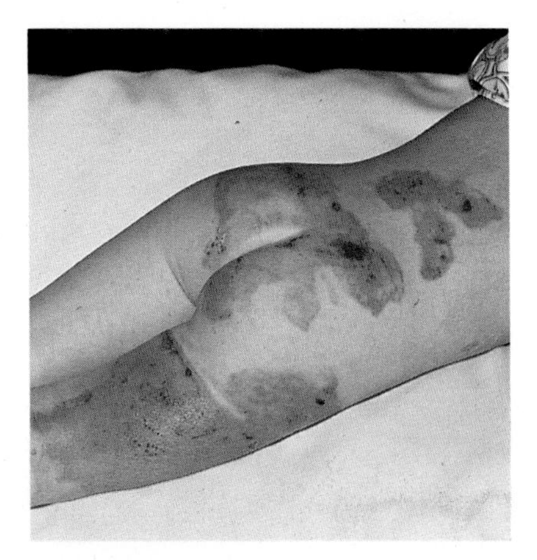

(c) At 16 days. Rapid, uncomplicated healing is seen with no painful dressing changes. No long term problems are anticipated.

dressings applied, or the wound must be surgically excised (Fig. 3.7). Moist surfaces encourage the growth of Gram negative bacteria such as *Pseudomonas aeruginosa*. However, the Gram positive cocci can tolerate dessication.

To encourage drying of the wound the patient should be nursed in a warm, dry atmosphere with free circulation of air. Sheets are not advisable as moist air is trapped against the wound. Limbs are elevated and the wound regularly washed with a topical antiseptic such as Hibiscrub or Betadine. Measures must be taken to avoid direct contamination by implementing stringent barrier nursing techniques, physical separation of the patients and an adequate flow of air around the wound.

The 'closed' method

By dressing the wound, contamination is minimized whilst providing an environment which will encourage epithelialization. Most dressings contain a topical antiseptic which also limits bacterial colonization. They should consist of a medicated, non-adherent layer in contact with the wound surface, several layers of cotton gauze to absorb any exudate, a thick layer of cotton wool to act as a further barrier and a

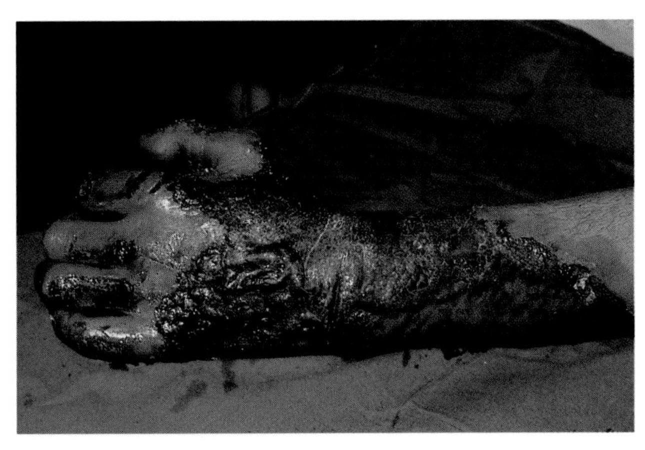

Fig. 3.4 Deep dermal burn. (a) This has been treated by exposure for two weeks.

(b) Under general anaesthesia, the eschar is lifted off.

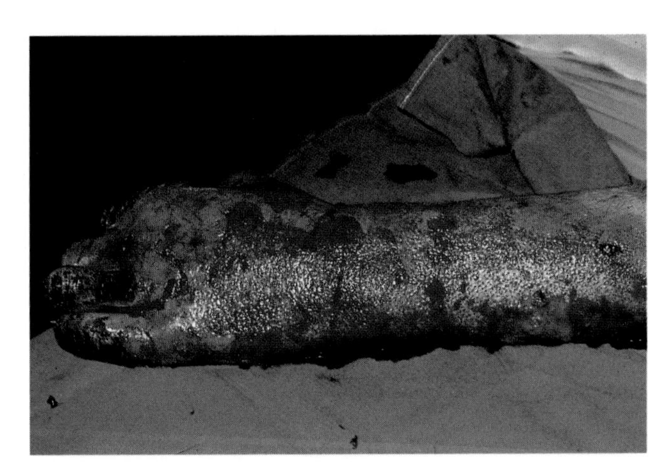

(c) A clean, healthy bed is revealed with no residual dead tissue. A thin split skin graft can be applied, and healing should be rapid and uneventful.

Fig. 3.5 Prevention of contamination. Although most of this eschar remains dry, some areas are becoming macerated and are therefore potentially contaminated. Thorough washing with an antiseptic soap solution, followed by drying in a very warm atmosphere, may allow the surface to dessicate and the exposure technique to be continued.

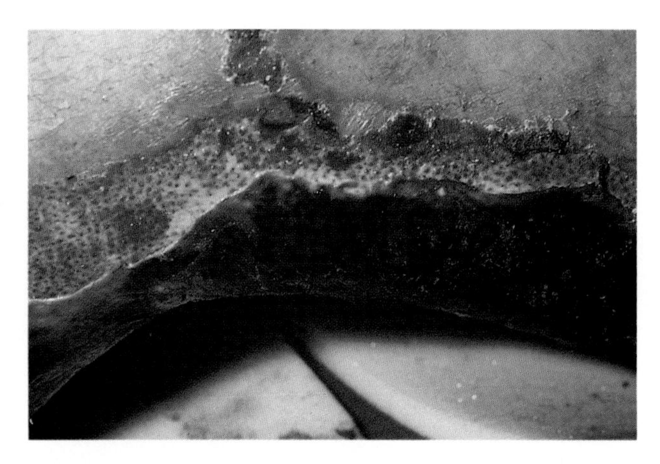

Fig. 3.6 Prevention of infection. The eschar is no longer dry and therefore no longer protects the underlying burn surface. The wound will be heavily colonized, and slight inflammation is noticeable at the edge. The exposure technique must be abandoned. Dressings may help to keep the number of colonizing organisms to manageable levels but do not eradicate them. Surgical excision should be considered.

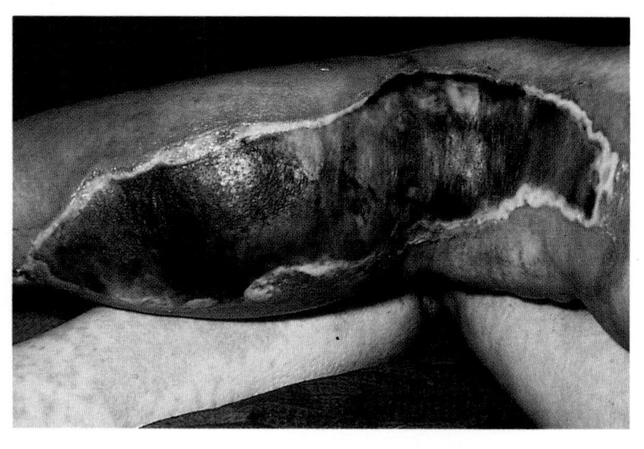

Fig. 3.7 Prevention of invasive sepsis. The red flare around this exposed eschar indicates local cellulitis followed closely by invasive sepsis. Infection in such an exposed wound is likely to be streptococcal and should be treated by high doses of penicillin, early wound excision and delayed grafting.

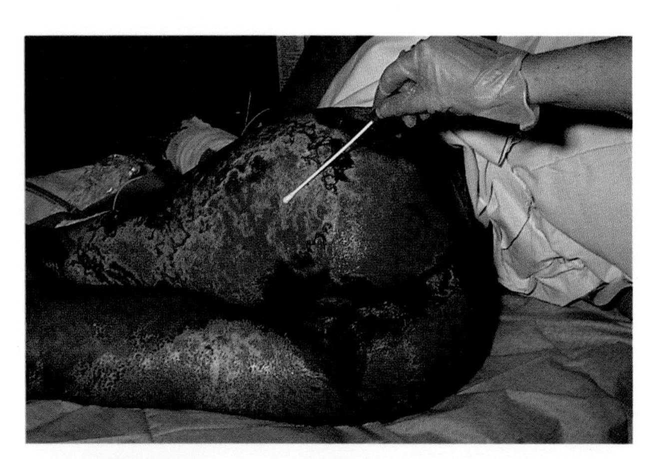

Fig. 3.8 Flamazine dressing. (a) Dirty dressings have been removed and wound swabs taken.

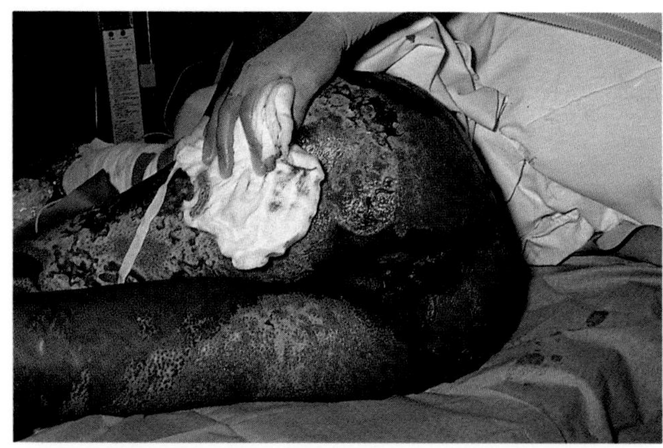

(b) The wound surface is cleaned.

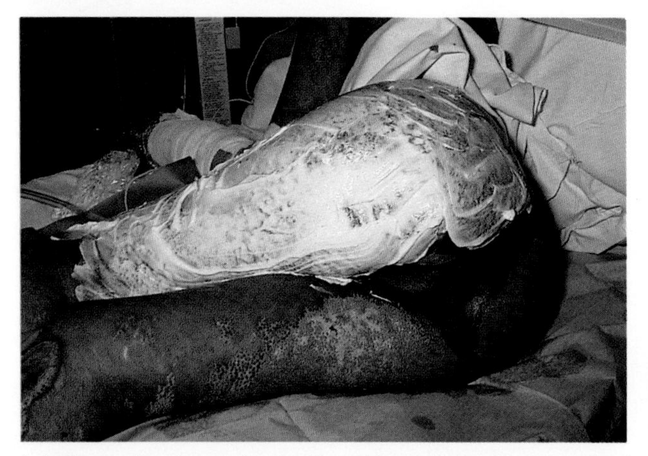

(c) Flamazine is applied with a gloved hand and worked uniformly into the wound surface to be dressed.

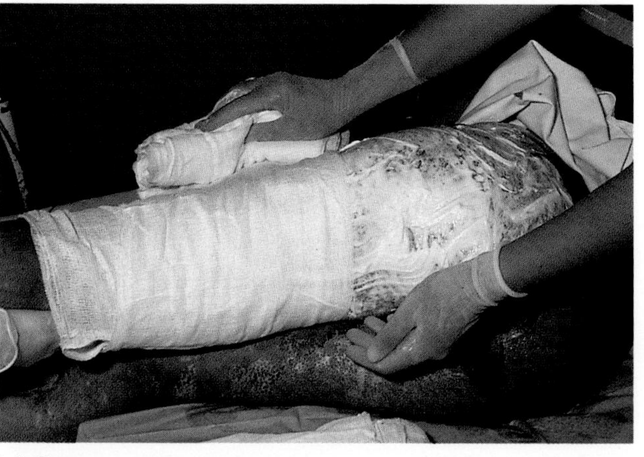

(d) Cotton gauze covers the cream.

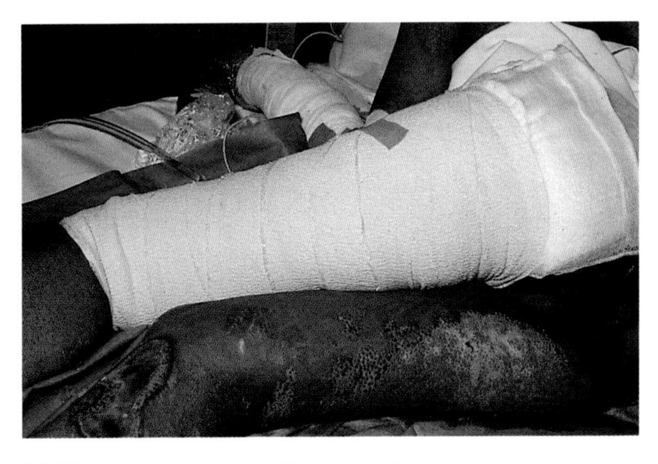

(e) Further cotton padding is held in place with crepe bandages.

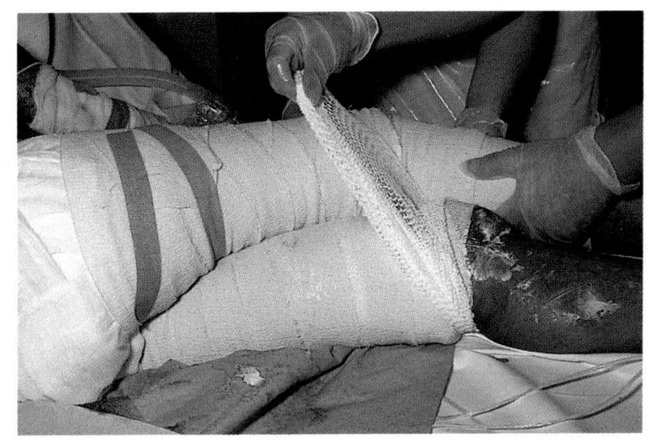

(f) Elastic net dressing is used to unite the leg dressings and to hold the gauze over the buttocks.

Table 3.2 Advantages and disadvantages of widely used antibacterial dressings

Dressing	Advantages	Disadvantages
Silver nitrate solution	Effective as prophylaxis against serious contamination	Ineffective in treating established infection Causes staining and leaching of chloride ions
Silver sulphadiazine cream	Wide antibacterial spectrum (Gram negative) Used widely as prophylactic dressing Eschar will dry after discontinuation	Delays spontaneous separation of eschar thus delays wound closure Does not penetrate eschar well Ineffective against established wound sepsis
Sulphamylon (Mafenide)	Penetrates eschar well Good antimicrobial action	May cause stinging pain which limits usefulness
Chlorhexidine Povidone iodine		Of limited use in larger burns
Nitrofurazone	Dries wound well Good antibacterial spectrum	May cause sensitivity rash
Bactroban (Mupirocin)	Very effective against Gram positive organisms (especially group A Streptococcus and Staphylococci	No effect on Gram negative bacteria

firm crepe bandage. The dressing should extend well beyond the area of the wound (Fig. 3.8). There is a range of dressings currently available; the most popular are listed in Table 3.2.

Wet dressings of 0.5% silver nitrate solution are very effective in preventing serious contamination but are ineffective in treating an established wound infection. They also stain skin and bed clothing a chocolate brown/black (Fig. 3.9). Care must be taken as silver nitrate leaches chloride ions from the circulation which must be replaced, thus careful monitoring of serum electrolytes is necessary. Silver sulphadiazine dressings are also widely used and are antibacterial against Gram negative organisms (Fig. 3.10).

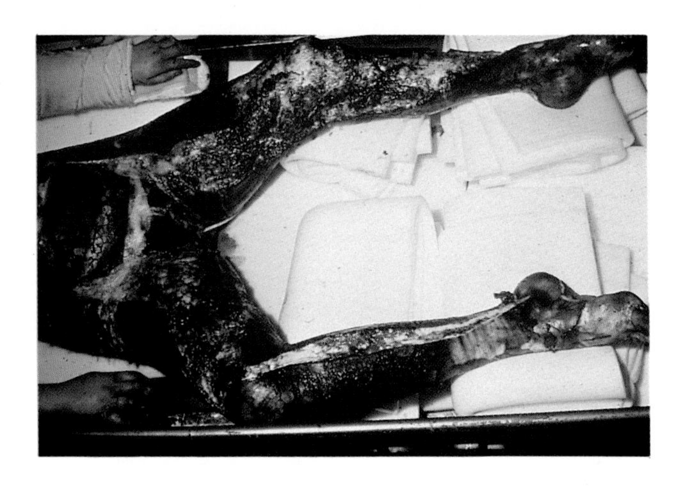

Fig. 3.9 A very deep burn. Treatment is with dressings of silver nitrate solution; the silver is precipitated out by protein and turns black on exposure to light.

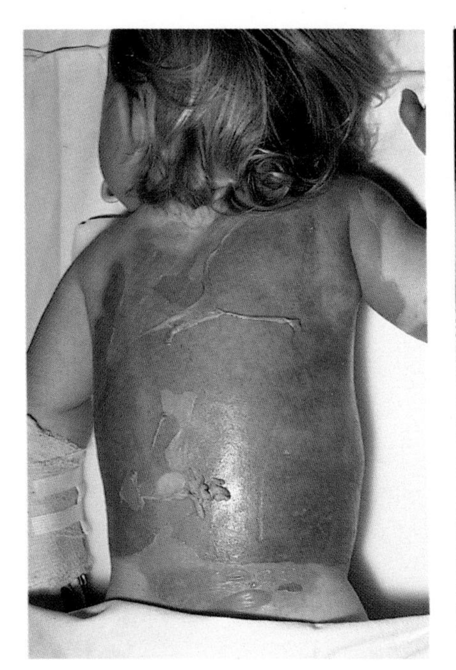

Fig. 3.10 An apparently superficial scald. (a) Treated in silver sulphadiazine cream.

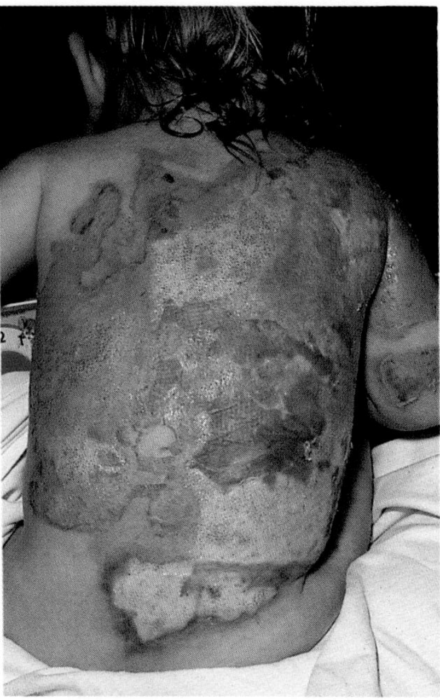

(b) At three weeks it is clear that the white exposed dermis over the right flank is deeper. The burn should be very superficially shaved, leaving as much dermis as possible, and a thin skin graft applied.

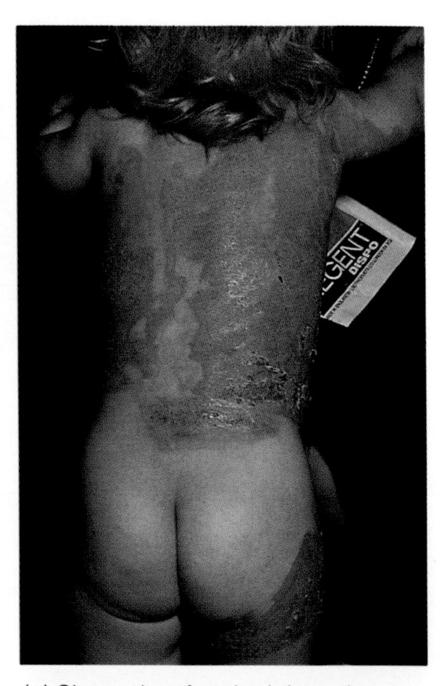

(c) Six weeks after the injury, the burn, grafts and donor site are healed.

Bacterial colonization of the wound cannot be prevented but frequent change of dressings will keep the absolute number of organisms down, thus preventing invasive sepsis. If infection does occur, the dressing must be changed at least once a day as it is the physical cleaning of the wound which is important. Alternatively, the wound may be excised and grafted (Fig. 3.11).

In the absence of infection, the wound can be left for several days as frequent dressing changes will expose the wound too often and encourage contamination, and will also damage the fragile epithelium thus delaying wound healing.

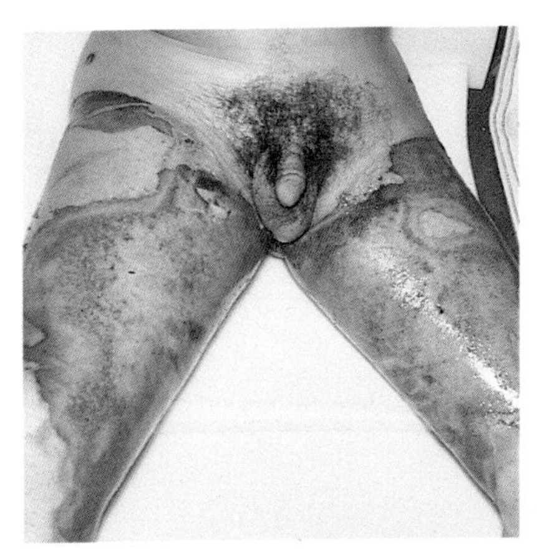

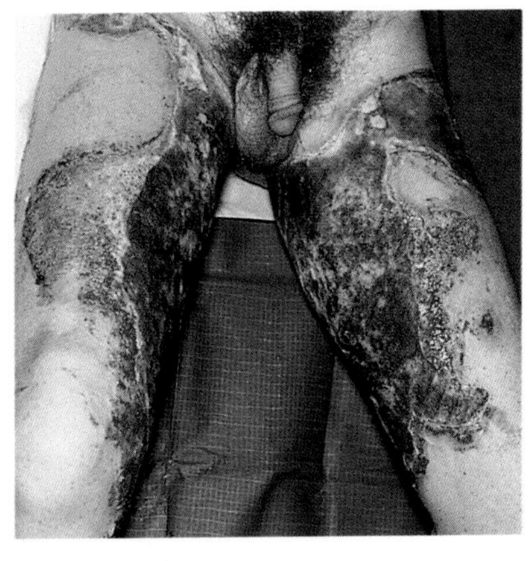

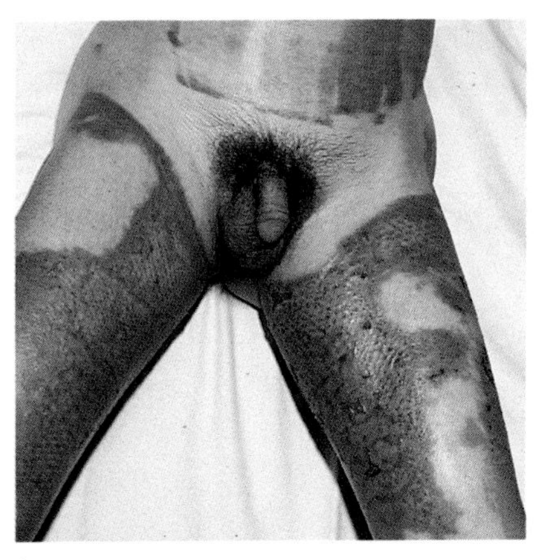

Fig. 3.11 Burns of mixed depth. (a) Treatment in dressings did not prevent significant colonization.

(b) At ten days there has been some separation of the eschar at the edges, but surgical excision at this stage would involve considerable loss of blood.

(c) The wound was grafted three weeks later with skin harvested from the abdominal wall, and is healed at six weeks after the injury.

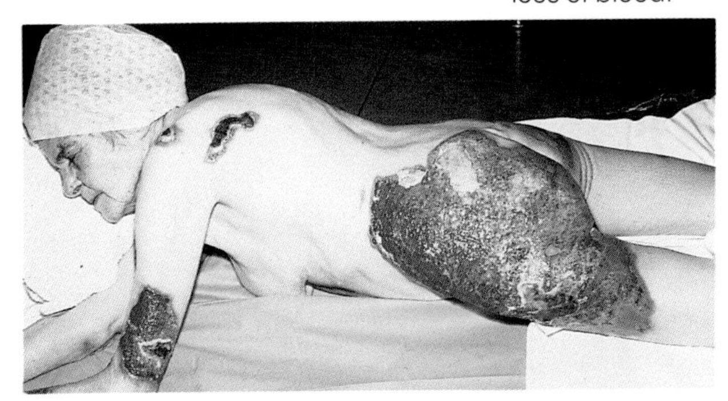

Fig. 3.12 Wound care in the elderly. (a) The eschar is allowed to separate spontaneously.

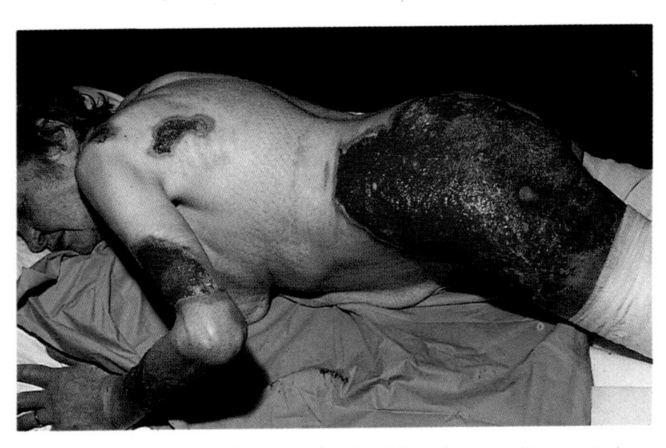

(b) Under a general anaesthetic thin skin grafts are taken and the wound gently scraped.

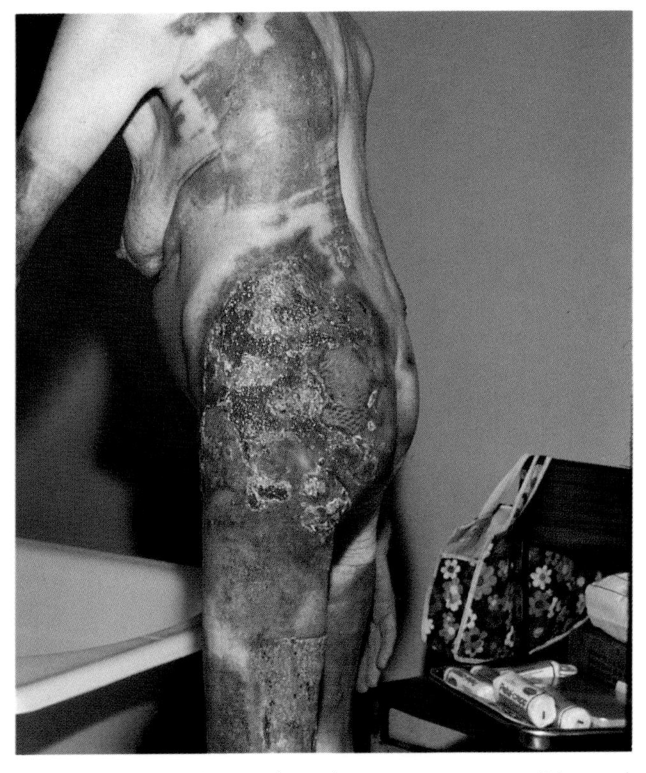

(c) Mobilization is continued as soon as possible and grafts are applied as a delayed procedure. In this case, healing is nearly complete at ten weeks.

In the elderly, treatment is directed towards daily baths and frequent change of medicated dressing, resulting in subsequent separation of the eschar (Fig. 3.12).

Dressing of superficial and dermal burns of the hand is carried out by applying antiseptic cream and covering with a loose-fitting plastic bag (Fig. 3.13). Contamination from the environment is prevented and the burns are surprisingly painless because they are kept macerated. Active movements are encouraged. The dressing should be changed daily and treatment continues for two or three weeks. If the wound fails to heal within this time, grafting should take place without delay.

The semi-open method

An alternative method is the semi-open/semi-closed technique, whereby antiseptic cream is applied over the burn surface but the wound is not formally dressed (Fig. 3.14). Reapplication of the cream at least once a day after washing is necessary and full move-

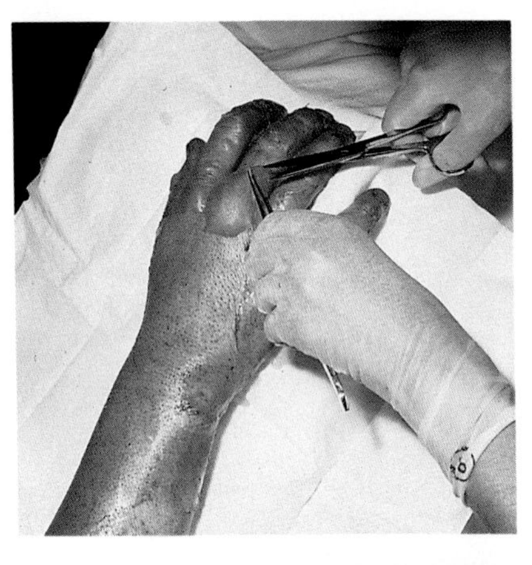

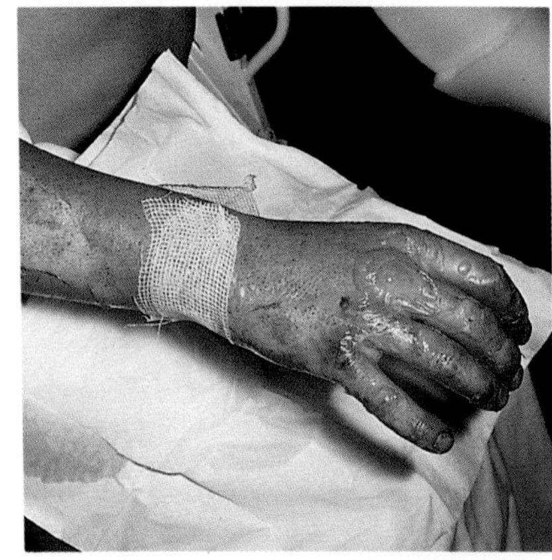

Fig. 3.13 Burns of the hand. (a) All blisters are emptied.
(b) Vaseline gauze is applied at the wrist to act as a barrier.

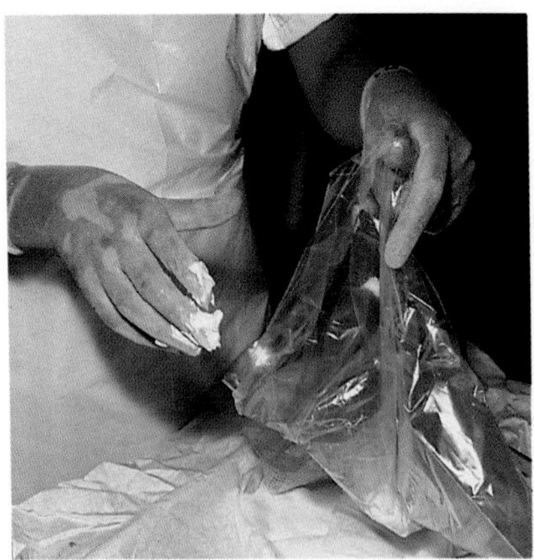

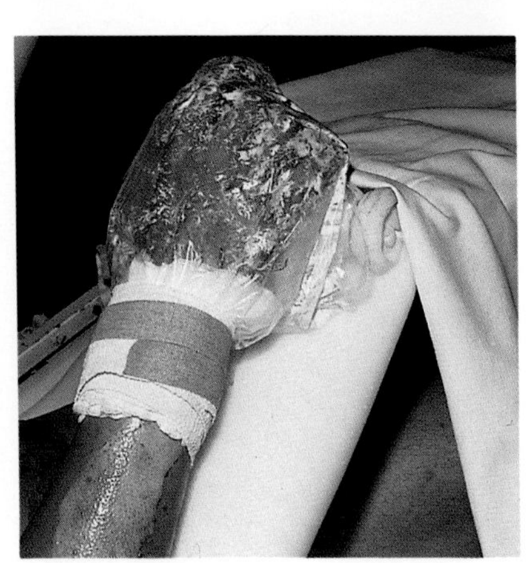

(c) Flamazine is placed on the hand and the plastic bag used as a cover.
(d) A light bandage is then applied to loosely hold the bag in position.

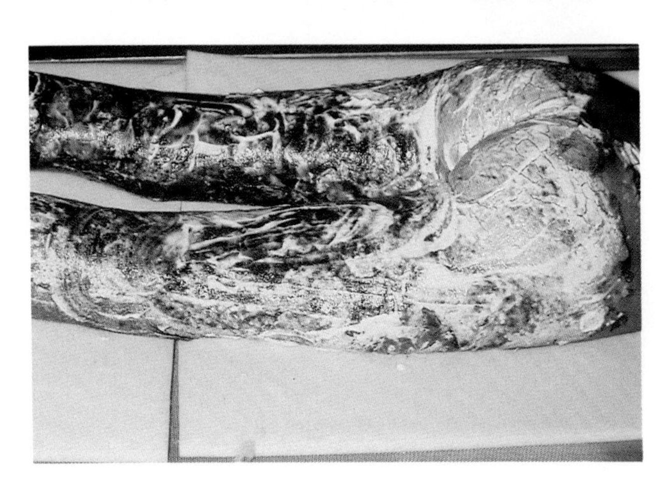

Fig. 3.14 Semi-open method. Silver sulphadiazine cream has been applied thickly to the burn, and should be washed off and reapplied at least once a day.

ments are encouraged. Good microbial prophylaxis is obtained, and if the technique is discontinued after 10 or 12 days and the eschar allowed to dry, the wound can be excised a few days later with relatively little blood loss. This method has the advantages of being quick and inexpensive.

ESCHAROTOMY

Deep circumferential burns that involve limbs or trunk are likely to contract, exerting a tourniquet effect as they dry (Fig. 3.15). Initially, venous return is impaired, or in the case of the chest wall, ventilatory excursion is limited. An incision must be made before this occurs and can be performed at the bedside (Fig. 3.16). The burns will be deep and therefore painless; an anaesthetic is not required although a bipolar coagulator is necessary to control bleeding. Fig. 3.17 shows the procedure of escharotomy of the hand and fingers, aiming to release tension.

Incisions should be made promptly as bleeding will be excessive if delay has led to vascular engorgement of the tissues, in which case a fasciotomy maybe indicated (Fig. 3.18). Furthermore, the escharotomy should provide adequate decompression of tissues. It can be seen in Fig. 3.19 that this has not occurred.

Initially, an incision is made in the centre of the burn down to fat and is extended along the line of the limb (Fig. 3.20). Pain indicates that the incision has reached an area of superficial burn, therefore further incision will not be necessary. Extension proximally and distally along the medial and lateral aspects of the limb releases the constriction (Fig. 3.21). Bleeding must be controlled with either a catgut stitch or a cautery and the incisions dressed and bandaged. Steps should be taken to avoid exposing tendons, vessels or nerves (Fig. 3.22).

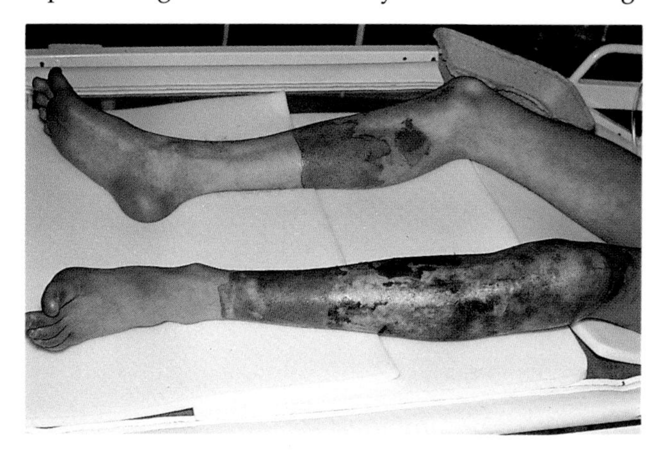

Fig. 3.15 Indication for escharotomy. A deep circumferential burn of the left leg is beginning to cause some obstruction to the circulation. The foot is slightly blue and an escharotomy should be carried out immediately. Such burn eschars are likely to contract, exerting a tourniquet effect.

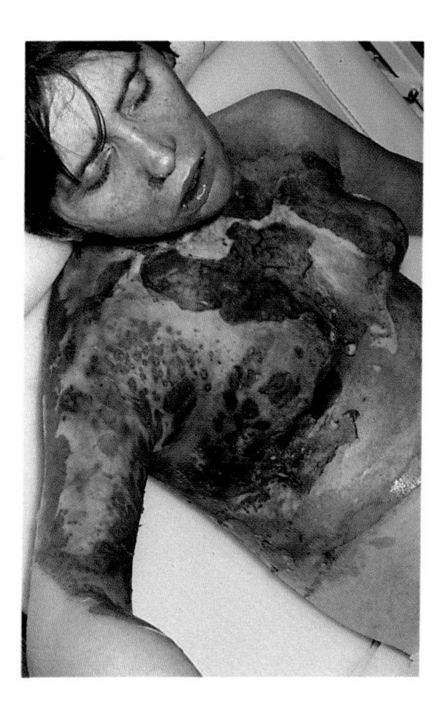

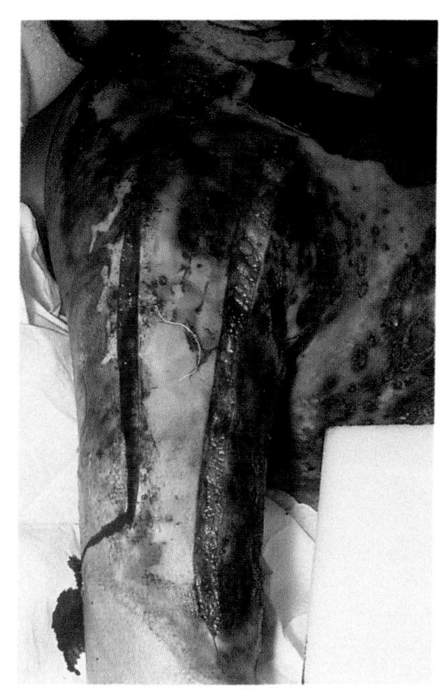

Fig. 3.16 Very deep flame burn of upper arm and chest wall. (a) These burns were exerting a constricting effect.
(b) Several incisions were required to relieve the congestion.

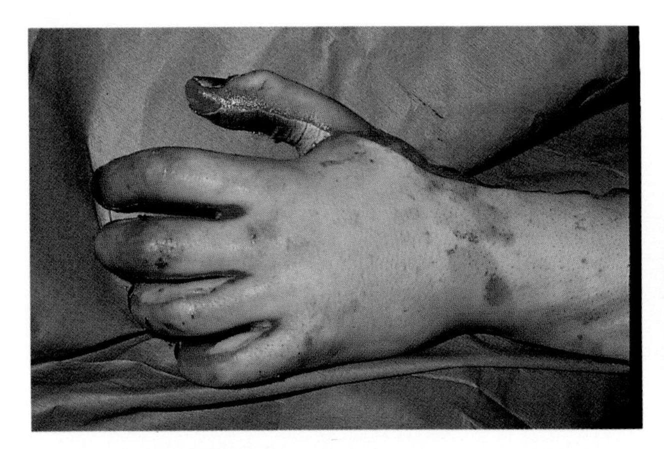

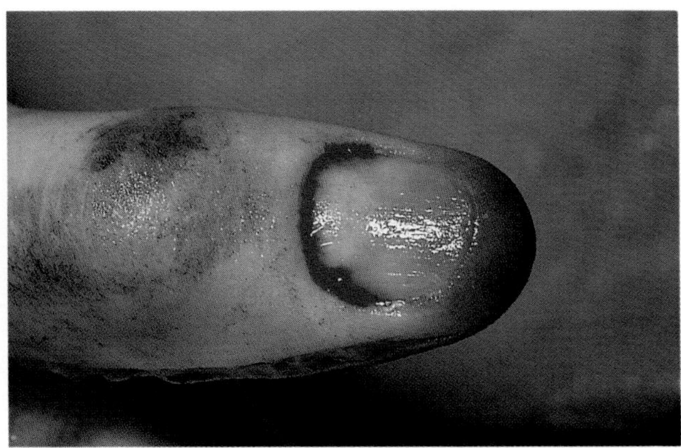

Fig. 3.17 Escharotomy of the hand. (a) A deep flame burn with early swelling of the dorsum. The position is characteristic, with extension of the metacarpophalangeal joints and flexion of the interphalangeal joints. A releasing incision has been made along the radial border of the hand, and further incisions are required to release the dorsum and fingers.

(b) Loss of the fingernail. This is seen in deep burns and implies that there has been substantial damage to the neurovascular bundles within the fingers, and that some loss of finger length is inevitable.

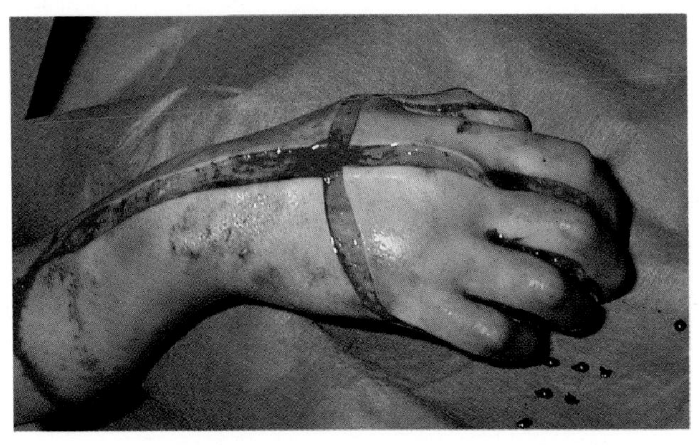

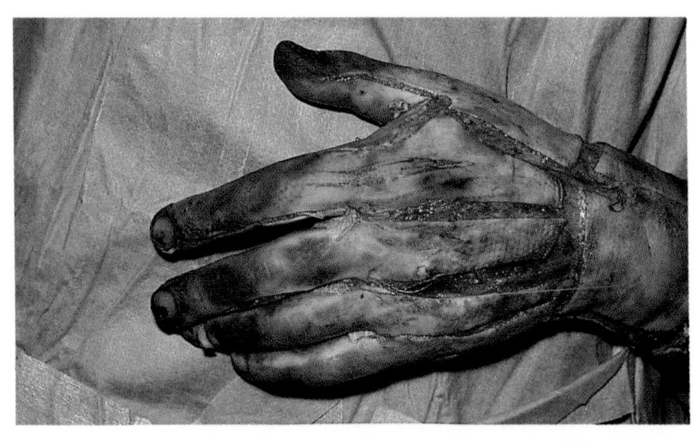

(c) Cruciate escharotomy. This incision, together with incisions made along the ulnar side of the fingers has been made to decompress the hand and wrist. The cutaneous nerves are less likely to be damaged and the incisions are largely unnoticed by the patient.

(d) Escharotomy of the hand. Thorough escharotomies release tension on both hand and fingers.

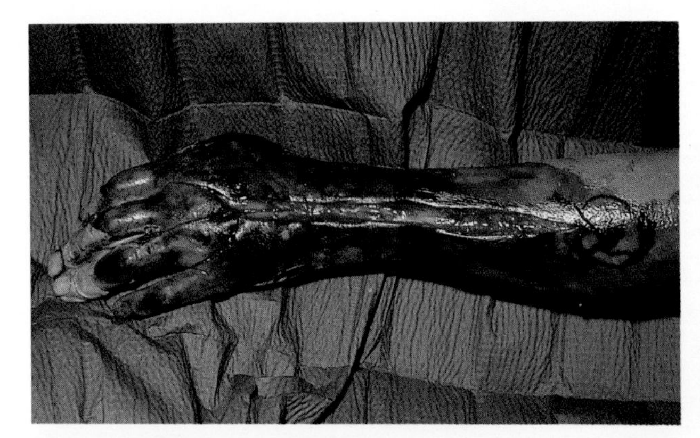

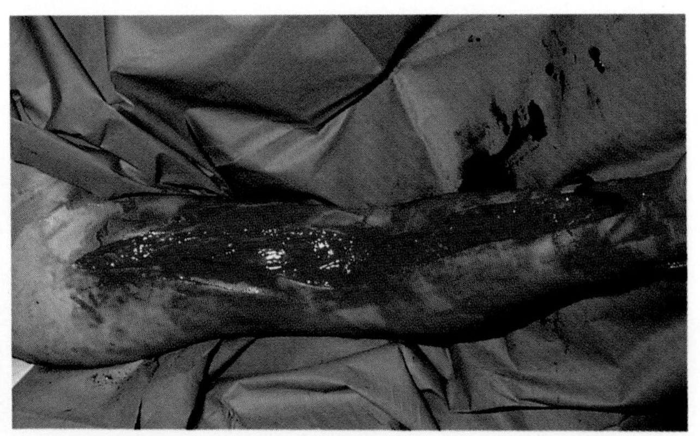

Fig. 3.18 Deep, charring burn. (a) This type of burn requires a thorough escharotomy and possibly a fasciotomy. Early wound excision is recommended.

(b) Escharotomy and fasciotomy of the arm. There was delay in performing the escharotomy and some muscle damage was anticipated. The fascia overlying muscle groups has been incised to allow for muscle swelling.

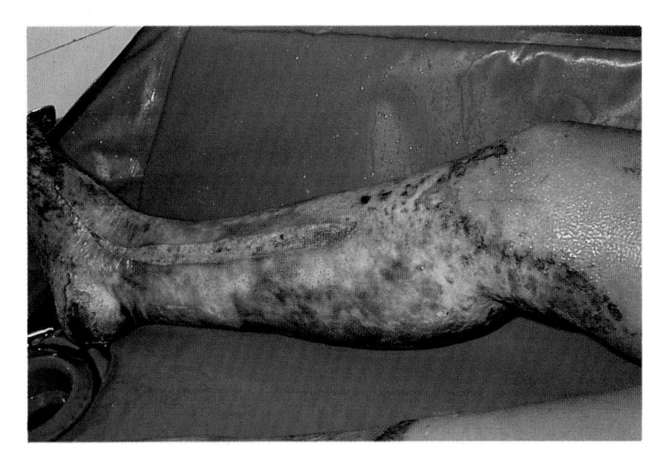

Fig. 3.19 An inadequate escharotomy. This was rather short and could not decompress the foot.

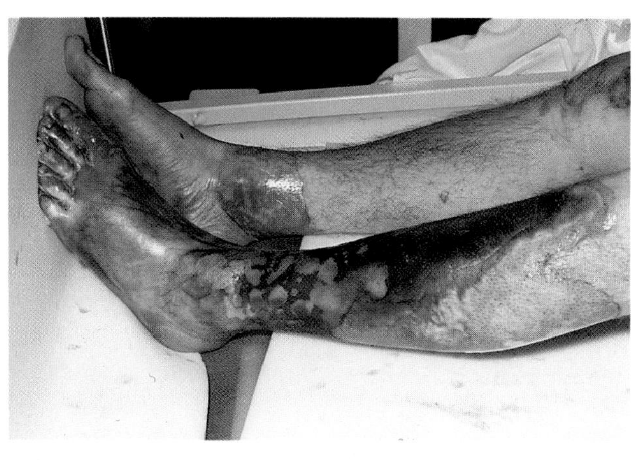

Fig. 3.20 A very deep burn to the lower leg. (a) There is obvious embarrassment to the circulation.

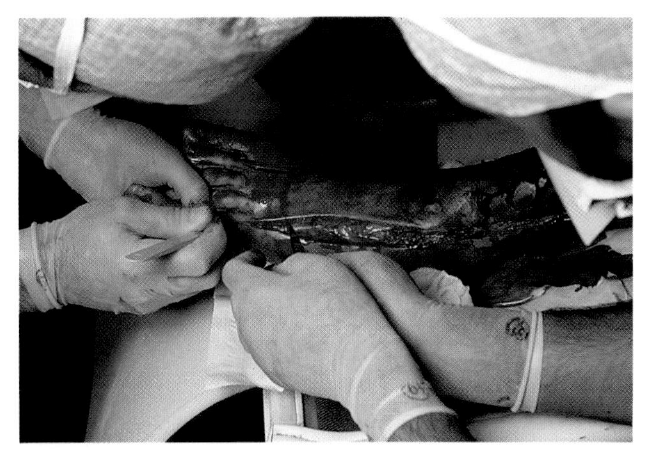

(b) An incision is done immediately at the bedside, with instant effect.

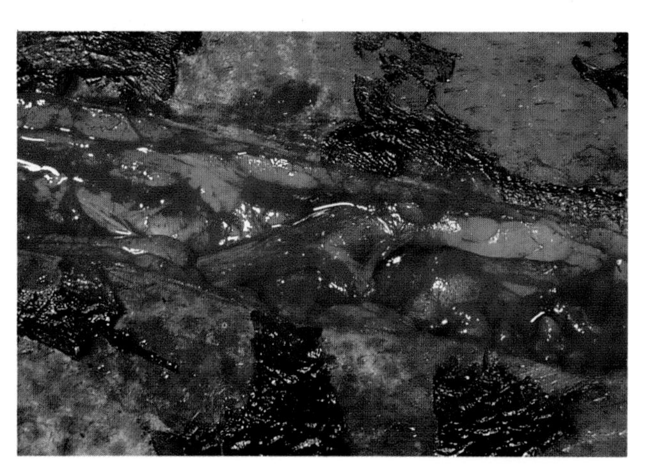

(c) There is no need to divide any vessels that bridge the incision once the tourniquet-like eschar has been released.

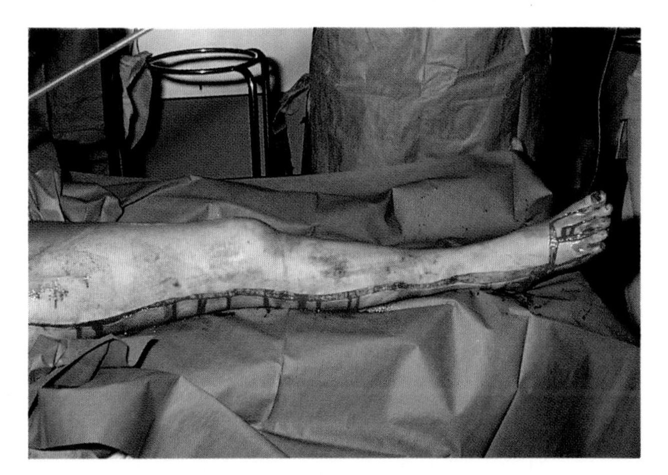

Fig. 3.21 Incisions to release constriction in deep burns. (a) A long releasing incision with extensions to the dorsum of the foot and toes. There is immediate release of the constriction with little immediate bleeding.

(b) Escharotomy performed behind the knees. Releasing incisions must be made through these very deep flame burns.

To perform a chest escharotomy, a chequer board pattern should be made with incisions running from the mid-sternal and mid-axillary lines downwards, with horizontal incisions running above and below the nipple, and if indicated, across the abdominal wall (Fig. 3.23). Bleeding is likely to be brisk and a blood transfusion will be necessary.

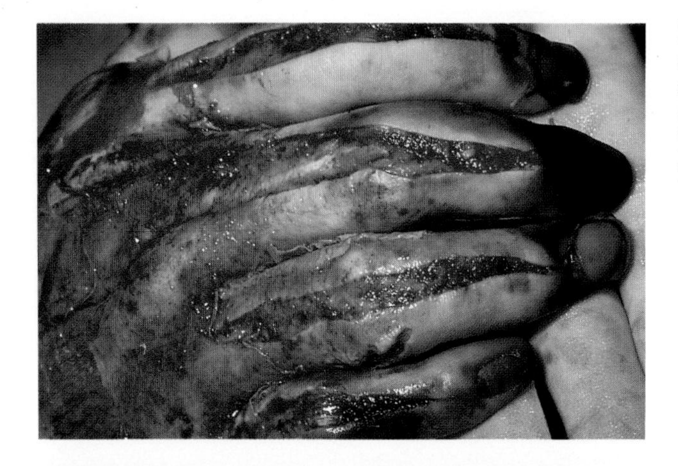

Fig. 3.22 Escharotomy of the hand. Incisions made along the dorsum of the fingers, as seen here, should be avoided as the delicate extensor tendons will be exposed and further damage will result. Flexion of the fingers will cause gaping of incisions and abrasion of tendons by the dressings.

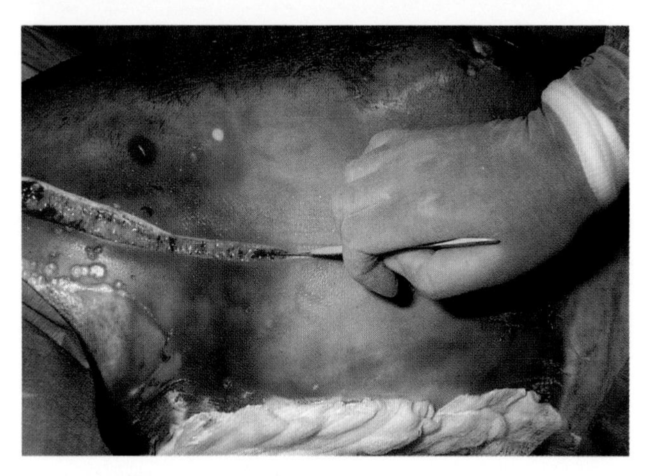

Fig. 3.23 Escharotomy of the chest. Escharotomies are made along the lateral chest wall with extensions to release respiratory excursions anteriorly. (a) The incision is initially almost bloodless.

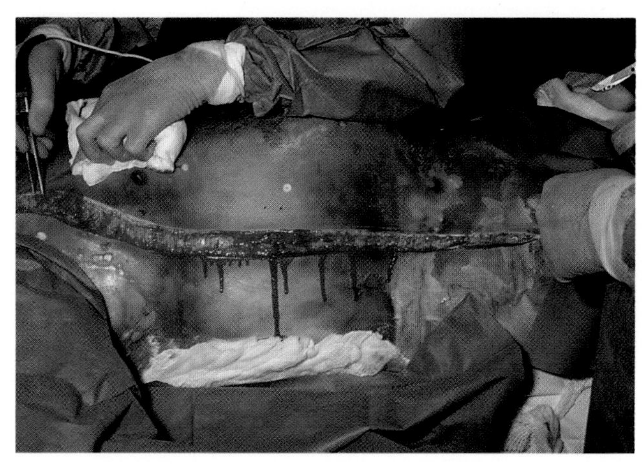

(b) A bipolar coagulator is used to control the haemorrhage after a few minutes.

(c) The incisions are dressed with ribbon gauze as an aid to haemostasis.

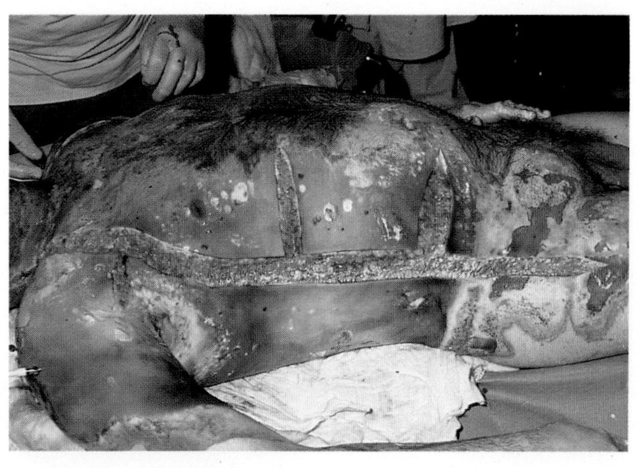

(d) Appearance three days later.

FIXATION OF FRACTURES

Fractures of the long bones can be safely held in reduction even when the skin over the fractures is deeply burned. If surgery is undertaken within 24 hours, direct open exposure of the fracture with reduction and immobilization using internal fixation is perfectly safe. Intramedullary fixation or plates can be used as the wound is not usually colonized by bacteria within 24 hours. Although it is important that the metalwork is covered with soft tissue, the incision through the burned tissue should not be repaired. A generous fasciotomy should be carried out and the wound left open to decompress the limb, with a skin graft used as cover.

When considering immobilization of a limb after 24 hours, it is likely that bacterial colonization of the wound will be significant and internal fixation should be avoided, and external fixation is advisable (Fig. 3.24). The reduction of the fracture is held with pins which can transgress the wound. Incisions should be made to allow tension-free introduction of the pins and should not be closed, thus assisting wound drainage. Perioperative antibiotic cover is not required and infective complications are rare, but may occur if wound excision is delayed (Fig. 3.25). If pintrack infection does occur, the pin can be removed and inserted elsewhere.

Fixation of fractures in the burnt patient is of great advantage. The fractures are made immediately painless and regular dressing changes are possible without an anaesthetic. They also heal rapidly and the patient can be mobilized quickly and without pain. Control of fractures using splints and plasters is inappropriate and should not be undertaken.

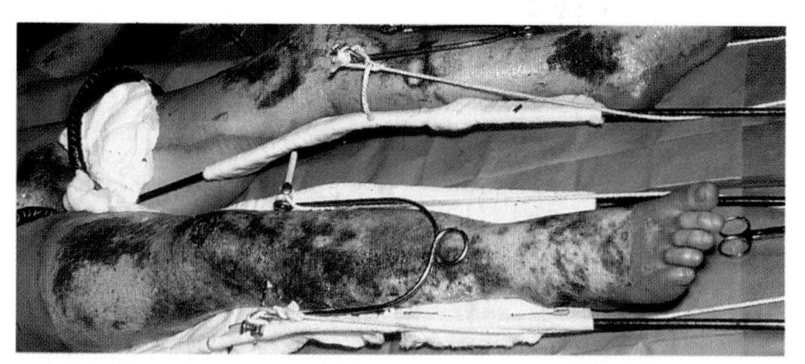

Fig. 3.24 External fixation of the femora. This patient, injured in an explosion, suffered extensive burns to the body. Beneath circumferential burns of the legs were fractures of both femora. (a) The fractures were initially held in Thomas's splints for transportation.

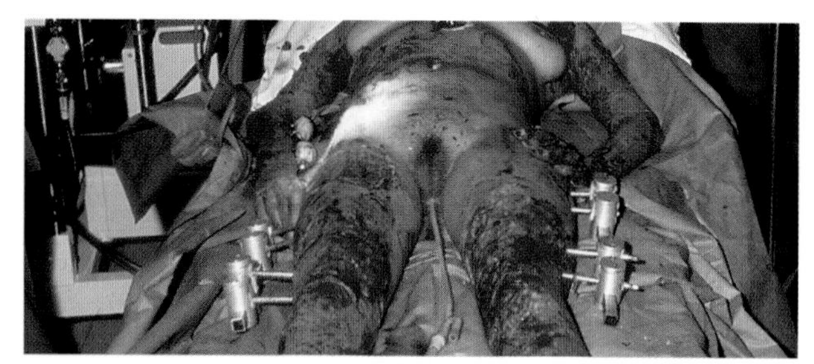

(b) External fixators were immediately applied to reduce and rigidly immobilize the fractures, which were instantly made painless.

(c) Repeated dressings and grafting were possible with little discomfort; there was no evidence of pintrack infection.

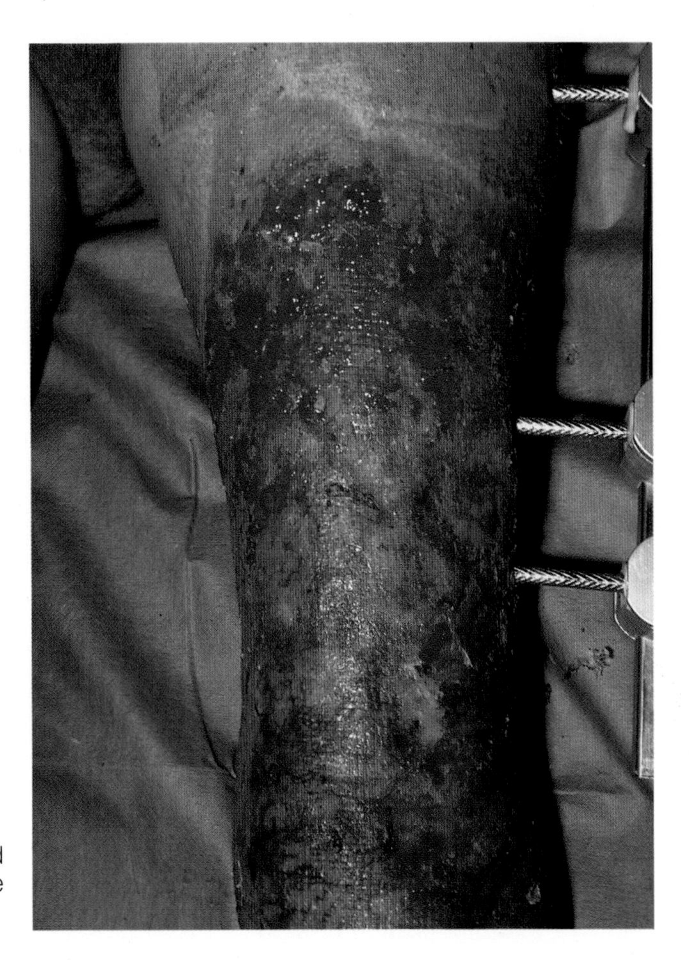

(d) Complete wound healing was achieved slowly, and the fractures united without delay or complication. The fixators were removed at 14 weeks.

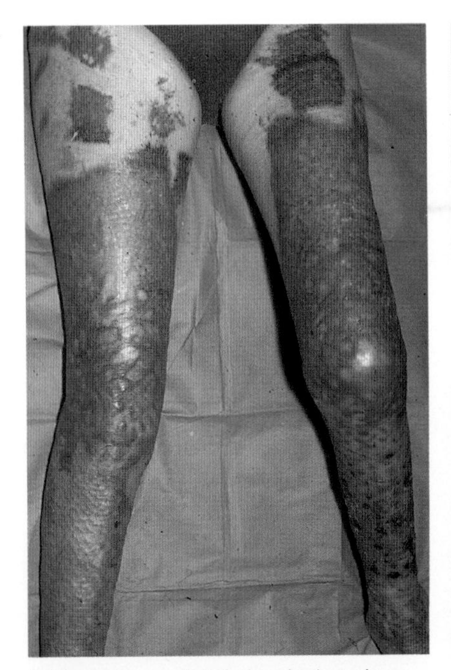

(e) Legs completely healed at six months, with no deformity and nearly normal knee flexion.

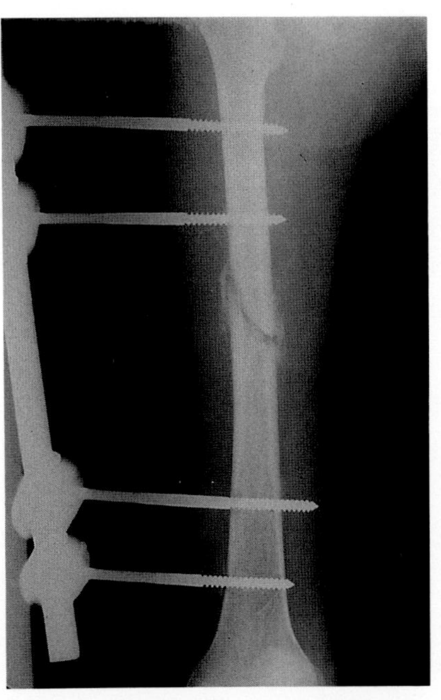

(f) X-ray of fractures at six weeks.

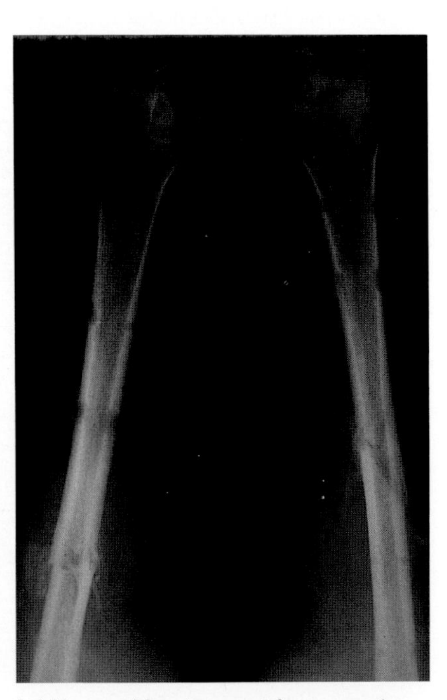

(g) X-ray of fractures at four months.

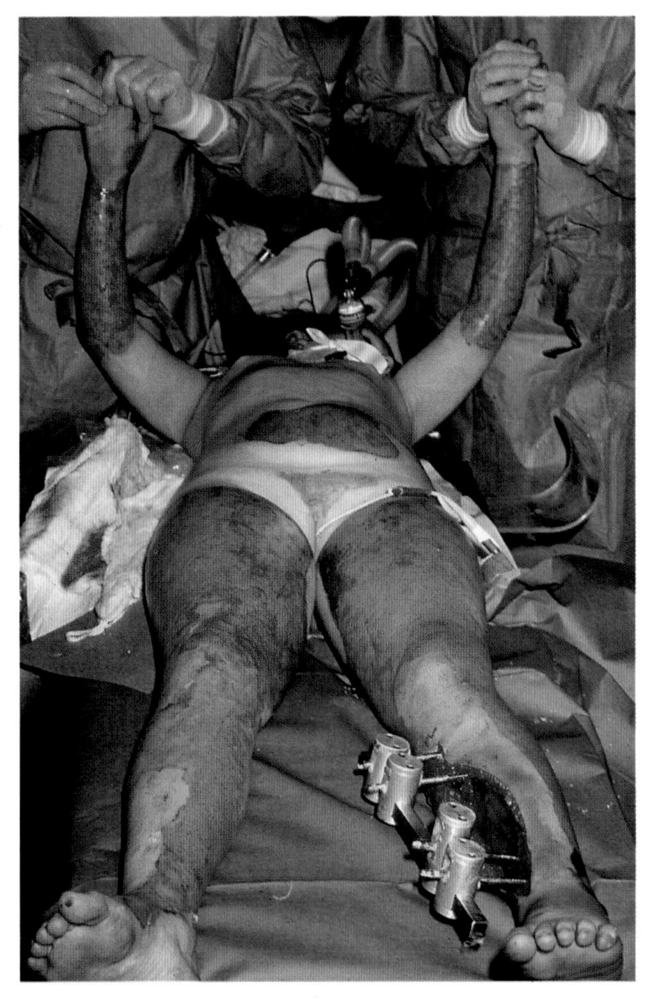

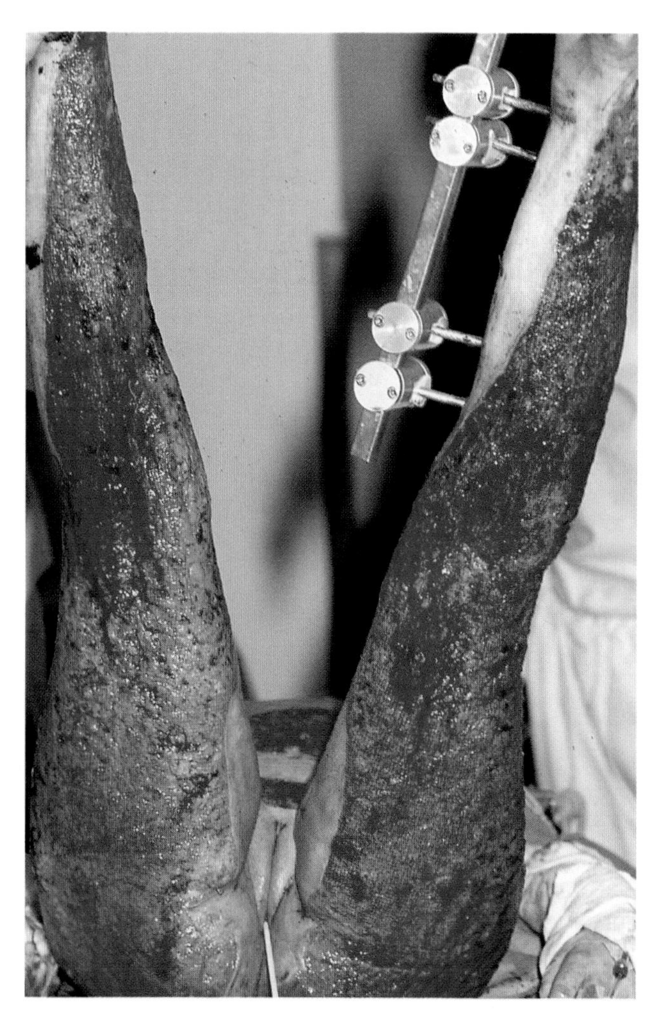

Fig. 3.25 External fixation with complications.
(a) Flame burns are seen together with a compound fracture of the left tibia. The fracture occluded the posterior tibial vessels and was immediately explored. A fragment was removed and a fasciotomy performed, the wound being left open.

(b) Excision of the burn was delayed and the wounds became heavily colonized with multi-resistant *Pseudomonas* organisms. Exposed bone was visible at the base but was covered by granulation tissue, enabling closure by simple skin grafting without the use of flaps.

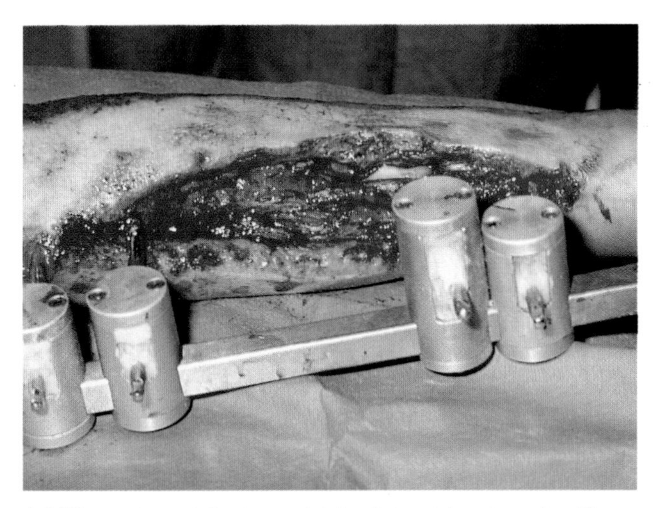

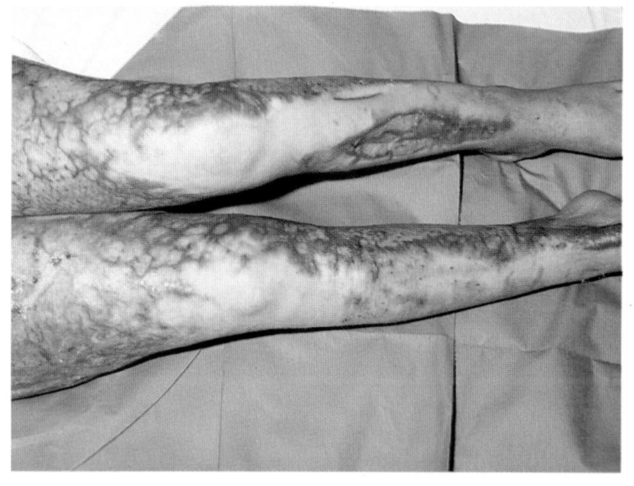

(c) The external fixator, which did not lead to significant infection, was removed at ten weeks, by which time the fasciotomy wound had healed.

(d) At 20 weeks the fracture is united, permitting full weight bearing. The burns are soundly healed.

4 Complications

INHALATIONAL INJURIES

The changes that occur in the lungs and airways following inhalation of smoke are complex and variable. Factors involved include the effects of heat, hypoxia, toxic products carried in the smoke and the secondary consequences of increased vulnerability of the lungs to fluid overload and infection.

Heat

Air that is hot enough to burn the skin will rarely damage the distal airways. Dry heat is dissipated rapidly and its direct effects are limited to the proximal airways and larynx. Swelling of the lips with oedema of the epiglottis and vocal cords, or stridor, signify impending upper airways obstruction and should be treated by urgent endotracheal intubation. Steam can carry up to 4000 times more heat than dry air, and its inhalation can cause extensive and irreversible damage to the distal airways and parenchyma of the lungs.

Hypoxia

The hypoxic component of an inhalational injury is of prime importance. Oxygen is consumed by combustion, thus lowering the amount available to the victim who, as a result, becomes confused. The presence of small amounts of carbon monoxide dramatically decreases the oxygen-carrying capacity of haemoglobin and the presence of hydrogen cyanide inhibits the cytochrome energy transfer system at cellular level.

Toxic products

Many toxic products are produced during combustion, particularly when the supply of oxygen is limited. The particles contained in smoke are mainly carbon, but toxic and irritant chemicals may also be present. Particles greater than 5 μm are trapped within the upper respiratory tract, but those of less than 1 μm will reach the alveoli. Chemical irritants may be water soluble, for example hydrogen chloride, ammonia and sulphur dioxide, and will have a rapid and corrosive effect on the mucosal surface. The effects of lipid soluble substances, for example aldehydes and the oxides of nitrogen, act on the endothelial membranes but may be slower in onset.

Secondary changes

The effects of bronchospasm, pulmonary oedema, loss of cilia, patchy atelectasis, retention of sputum and changes in endothelial permeability make the lungs vulnerable to secondary infection which will inevitably lead to a serious deterioration of lung function (Fig. 4.1). Prophylactic antibiotics are not of proven benefit, but appropriate antibiotics should be administered after frequent analysis of the sputum and wound bacteriology.

Exposure to carbon monoxide

Carbon monoxide impairs oxygen transport due to its greater affinity for haemoglobin (0.1% carbon monoxide in inspired air will bind with 62% of the

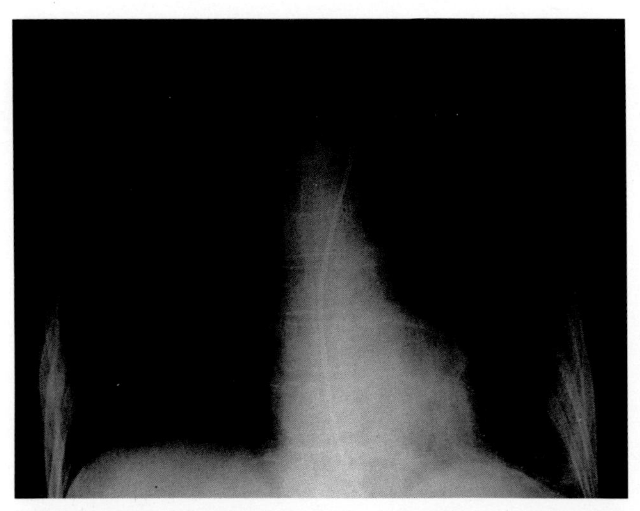

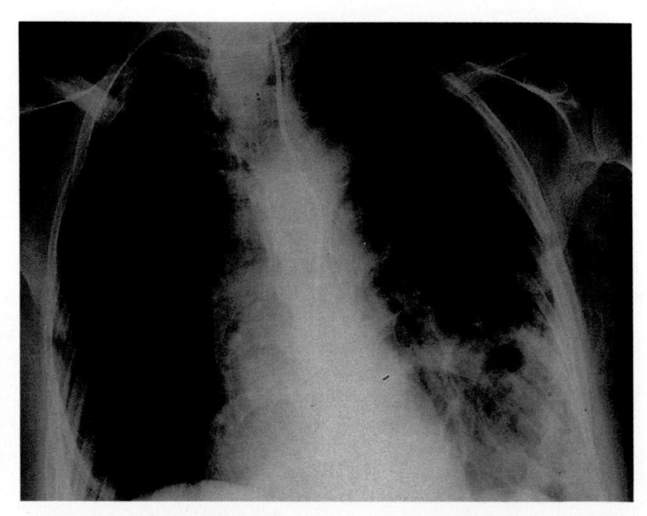

Fig. 4.1 Deterioration of lung function. (a) Development of left lower lobe pneumonia. Suitable antibiotics are needed. (b) An aspirate of the left lower lobe secretions, via a bronchoscope, may be diagnostically helpful.

available haemoglobin). Furthermore, by shifting the oxygen dissociation curve to the left, oxygen bound to the haemoglobin is less readily unloaded in the tissues. Carbon monoxide also competes with the cytochrome oxidase system at cellular level, further limiting oxygen utilization.

Diagnosis
Most hospitals are not equipped to measure carboxyhaemoglobin levels, therefore diagnosis of significant exposure to carbon monoxide will be imprecise. A range of symptoms related to level of carboxyhaemoglobin are listed in Table 4.1. The arterial blood gas estimation, p_aO_2, which measures the oxygen tension (amount of oxygen dissolved in the blood rather than the amount combined with haemoglobin) may be normal, therefore the history of the accident will be an important diagnostic factor. Oxygen saturation is a more accurate assessment, and a low oxygen tension in mixed venous blood will imply carbon monoxide poisoning.

Table 4.1 Effects of increasing levels of carboxyhaemoglobin

Carboxyhaemoglobin (%)	Symptoms
20	Severe headaches
30	Headache, irritability, indistinct vision
50	Confusion, collapse
60	Convulsions, unconsciousness

Treatment
Oxygen should be administered through a high flow face mask as soon as possible (Fig. 4.2). In young children, administration of oxygen in a croup tent may be more appropriate (Fig. 4.3).

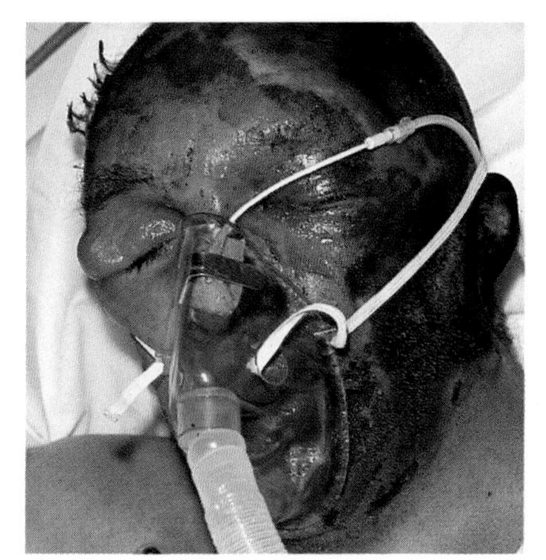

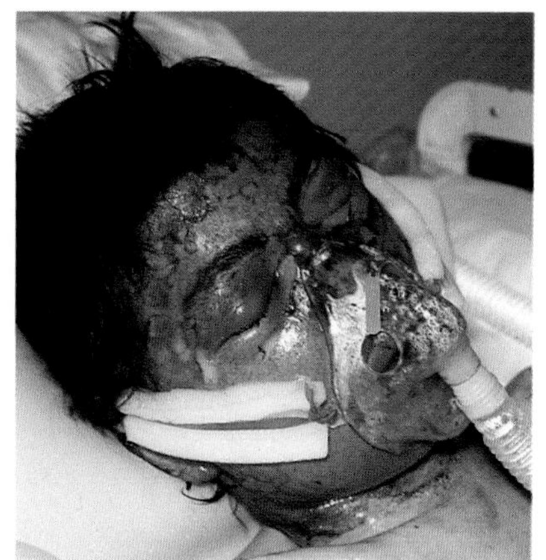

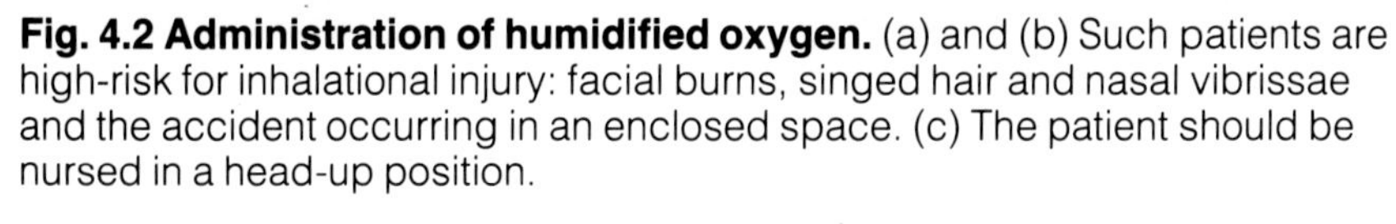

Fig. 4.2 Administration of humidified oxygen. (a) and (b) Such patients are high-risk for inhalational injury: facial burns, singed hair and nasal vibrissae and the accident occurring in an enclosed space. (c) The patient should be nursed in a head-up position.

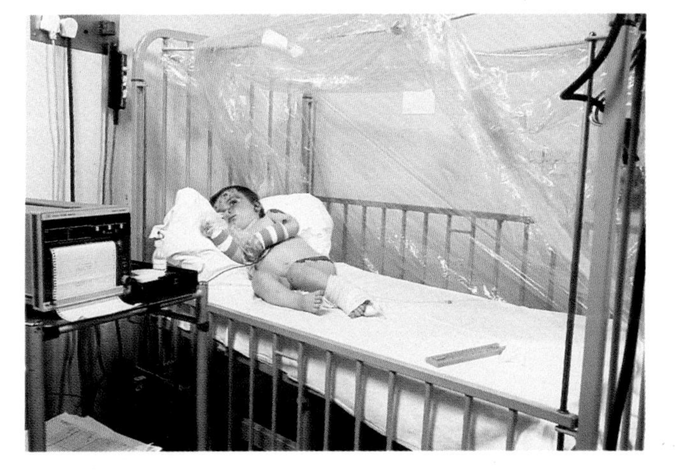

In the presence of significant neurological or electrocardiographic signs the patient should be ventilated with 100% oxygen, or placed in a hyperbaric chamber at 2.5 atmospheres if this is available. In the presence of severe neurological signs, it is recommended that ventilation is combined with hypothermia to 30°F, high dose steroids, measures to prevent hypothermia and limitation of fluids. It may be necessary to continue this treatment for two to three days.

Fig. 4.3 Use of a croup tent. As it is often difficult to fit a face mask to a child, this is an excellent alternative. Accurate humidification can be maintained and the child observed and comforted without difficulty.

Clinical presentation in inhalational injuries

Patients may present a range of symptoms from mild upper respiratory tract irritation to rapid death. More significantly, the onset of symptoms may be delayed for 24 to 48 hours.

During this initial phase (up to 6 hours), there may be few signs of injury. Over the next 48 hours, the patient may develop evidence of airways obstruction with wheezing, coughing, copious sputum, stridor and hoarseness, or pulmonary oedema (Fig. 4.4) and increasing signs of parenchymal damage. After 48 hours, in patients with significant inhalational injuries there will be a progressive bronchopneumonia leading to respiratory failure (Fig. 4.5). This development of serious necrotizing lung damage is frequently seen, particularly with extensive burns, and appears to be related to immunosuppression. Mortality is high and subsequent surgery places a further load on the lungs, which can prove fatal.

However, if the infection is not too severe, and this phase is overcome, there will be a prolonged period of gradual resolution, often incomplete, leading to the emergence of chronic lung disease (Fig. 4.6). This condition is exacerbated by extensive cutaneous burns.

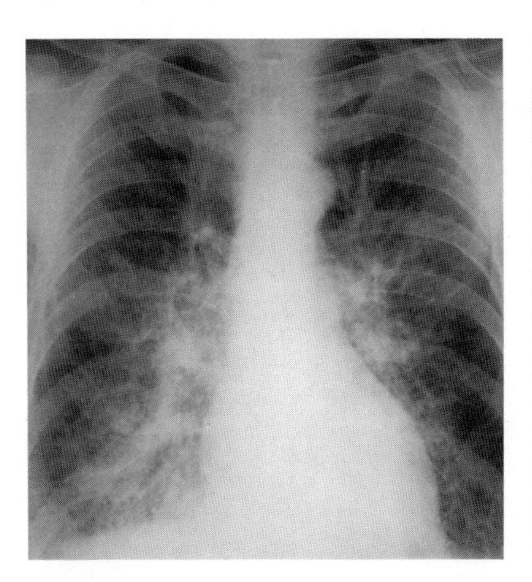

Fig. 4.4 X-ray of acute pulmonary oedema. There is rarely any abnormal appearance when the radiograph is taken shortly after exposure to smoke.

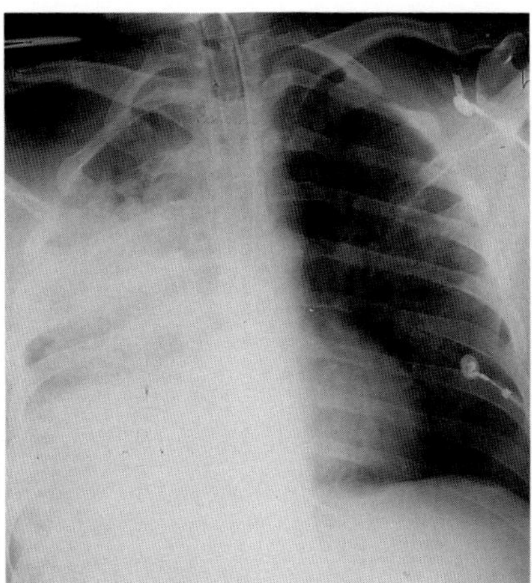

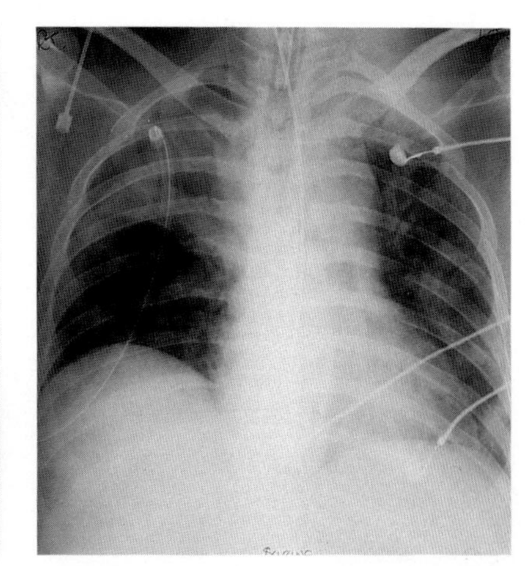

Fig. 4.5 Inhalational injury. (a) Injuries on presentation. (b) Appearances 24 hours later. Collapse of right upper lobe with pneumothorax, chest drain and signs of left sided bronchopneumonia.

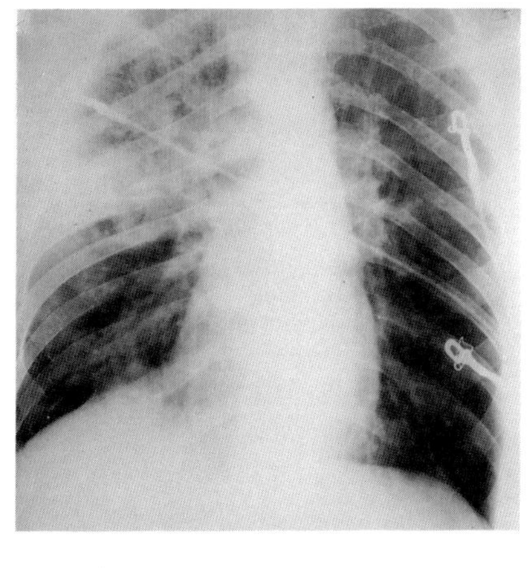

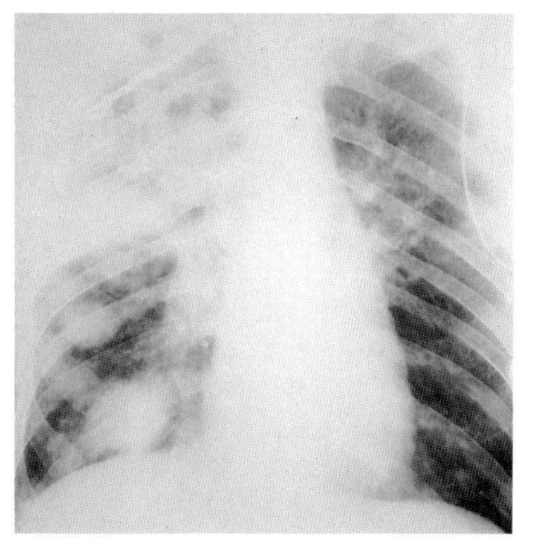

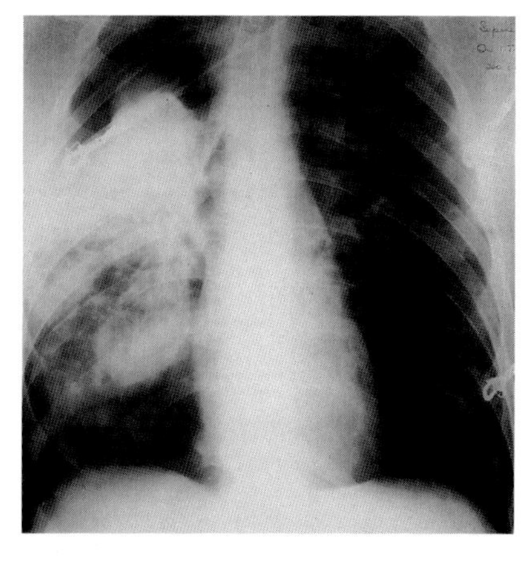

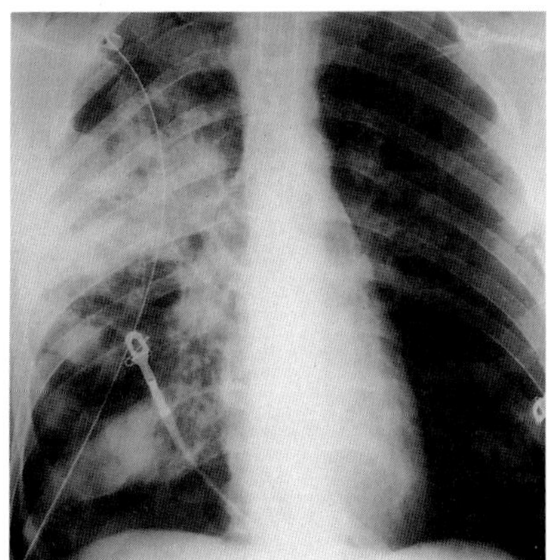

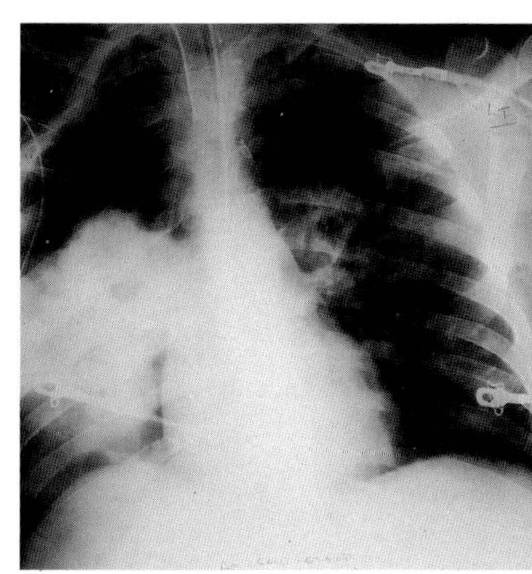

Fig. 4.6 Inhalational injury. Rapid development of signs of right upper lobe collapse, followed by consolidation. The patient was on a ventilator and receiving PEEP. (a) Partial collapse of right middle lobe.
(b) This is followed by extensive bronchopneumonia.
(c) Pneumothorax with almost total collapse of lung.
(d) Apical chest drain with some re-expansion.
(e) Further collapse with subcutaneous surgical emphysema.

Pathology

(i) The proximal airways show oedema with loss of cilia.
(ii) The distal airways show constriction of the bronchi, with the presence of mucosal sloughs and retention of sputum.
(iii) The parenchyma shows loss of surfactant, increased permeability of the alveolar membrane (due to the inhalational injury and the burn injury itself), pulmonary oedema and patchy atelectasis, secondary infection and extensive bronchopneumonia.

Diagnosis

(i) History of the accident, where available.
(ii) Physical examination, determining any alterations in the level of consciousness, presence of hoarseness and stridor, burns to the face and pharynx, oedema of the lips, and copious sputum containing smoke particles (Fig. 4.7). Bronchoscopy is of little diagnostic value.

Fig. 4.7 Carbon stained pharyngeal aspirate. This follows smoke inhalation and should be carefully monitored.

(iii) Investigations with careful and repeated monitoring, including p_aO_2, metabolic acidosis and oxygen saturation, and chest X-ray.

Treatment

(i) Humidified oxygen via face mask, which promotes rapid healing (Fig. 4.8), and sedation.

(ii) In the case of stridor, endotracheal intubation is needed, preferably via the nasal route (Fig. 4.9), although intubation via the oral route may be necessary (Fig. 4.10). Advantages and hazards of endotracheal intubation are listed in Table 4.2.

(iii) Physiotherapy, postural drainage and suction.

(iv) Nursing in a head up position.

(v) Administration of bronchodilators – aminophylline and salbutamol.

(vi) Ventilation with PEEP, if necessary.

(vii) Therapeutic bronchoscopy.

(viii) Specific antibiotics where sensitivities are available.

Table 4.2 Advantages and hazards of endotracheal intubation

Advantages
Control of access to airways
Bronchoscopy and suction
Overcome ventilatory work done
Decrease stress and anxiety
Hazards
Rapid contamination of the airways with the organisms growing on the burn wound
In physiotherapy, hypoxic bradycardia and cardiac arrest are a risk: avoid with preoxygenation
Oxygen toxicity and its effects on the lung
Barotrauma: pneumothorax and surgical emphysaema especially with PEEP
Lung abscess and emphysaema
Hazardous to perform if not experienced.

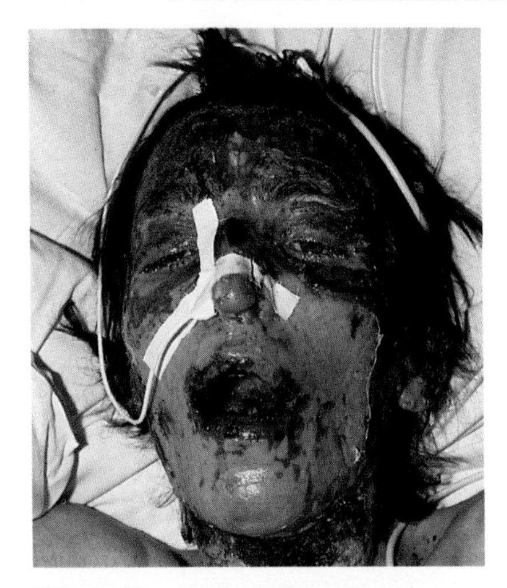

Fig. 4.8 Rapid healing of facial burns. These have been covered with a mask conveying humidified oxygen.

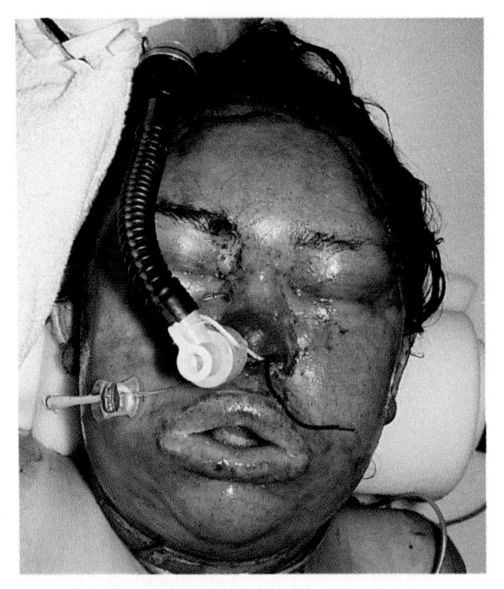

Fig. 4.9 Endotracheal intubation via the nasal route. This is the ideal route: well-tolerated, therefore needing less sedation, and easier to maintain in position.

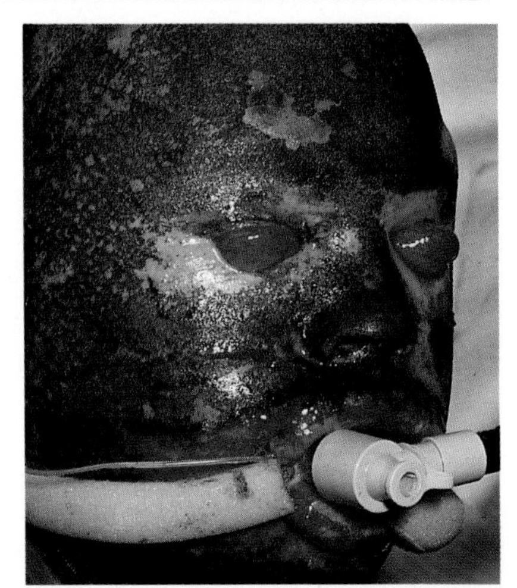

Fig. 4.10 Endotracheal intubation via the oral route. Although the nasal route is preferable, burns to the nares, mucosal oedema or septal deviation may make this difficult. An oral tube is more easily displaced and has to be tied very carefully into place. Light foam protection helps to prevent damage to the burned tissue by tapes.

TRACHEOSTOMY

There is rarely an indication for emergency tracheostomy, as endotracheal intubation or fibreoptic laryngoscopy can be carried out by an experienced anaesthetist. However, in deep burns of the neck this method becomes increasingly more difficult with neck flexion, thus early tracheostomy is often indicated (Fig. 4.11). As the incisions are through dead tissue, the wounds are rapidly contaminated, often resulting in severe complications. After two or three weeks of endotracheal intubation, movement of the tube, the pooling of secretions and the effects of pressure from the cuff become detrimental, and granulations start to develop. Further damage can be prevented by performing a tracheostomy. Additionally, if severe pulmonary infection has developed and there is a need for frequent suction, a tracheostomy allows the patient to be weaned from the ventilator. This method is advantageous in that heavy sedation or paralysis is not necessary and the patient can co-operate readily with the physiotherapist. Bronchial suction and lavage are straightforward and it is possible to begin mobilization at this stage. Figs. 4.12–4.17 show the sequential steps in performing a tracheostomy.

A Bjork flap tracheostomy is recommended as the flap of the tracheal wall stitched to the skin facilitates reinsertion of the displaced tube. One theoretical disadvantage of this method is tracheal stenosis, but this is compensated for by the extra degree of safety. Other complications associated with tracheostomy include bleeding, invasive infection and necrosis, and mediastinitis.

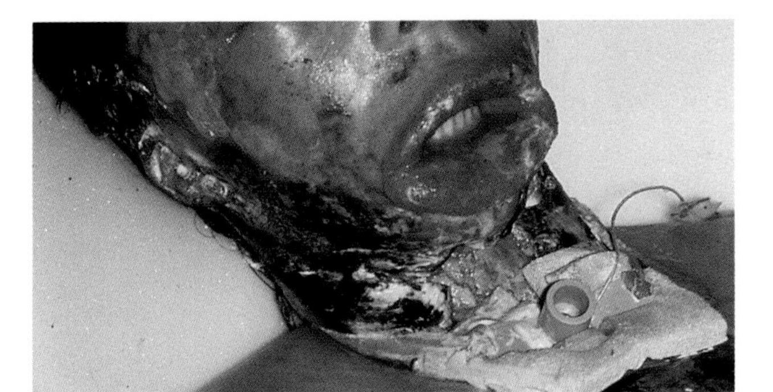

Fig. 4.11 Indication for early tracheostomy. The burns of the neck are very deep, with obvious oedema of the face and lips.

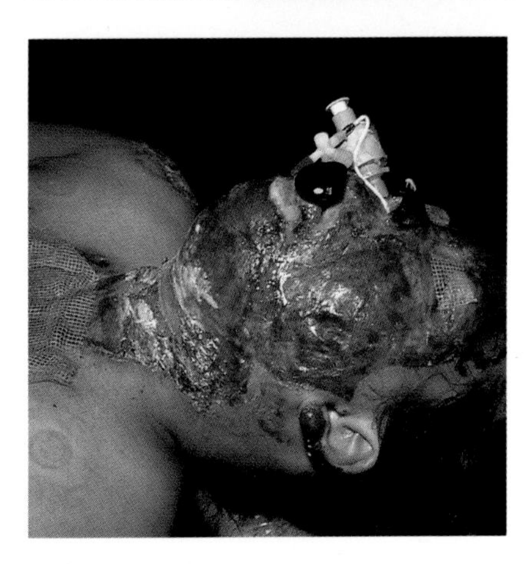

Fig. 4.12 Preparation for tracheostomy. With the nasoendotracheal tube in situ, the neck was extended and a sand bag placed beneath the shoulders. The skin was cleaned and draped, and after consultation with the anaesthetist a range of suitably sized tracheostomy tubes were assembled. These tubes were checked, their cuffs tested and connecting tubes and adaptors fitted.

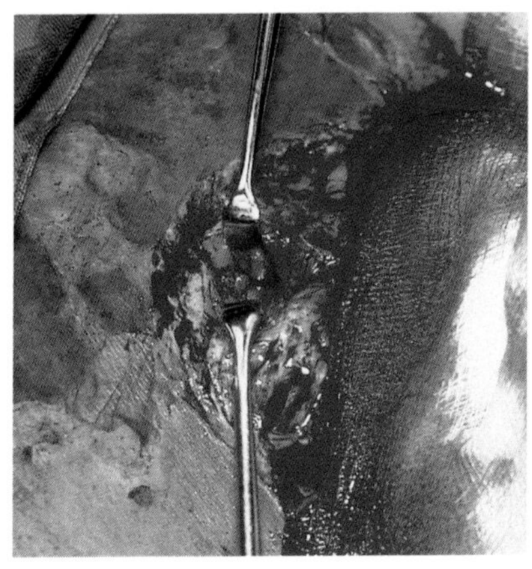

Fig. 4.13 Excision of burned skin. (a) The burned skin low in the neck, over the site of proposed tracheostomy, was generously excised down to healthy fat.

(b) The strap muscles are separated and held apart by retractors.

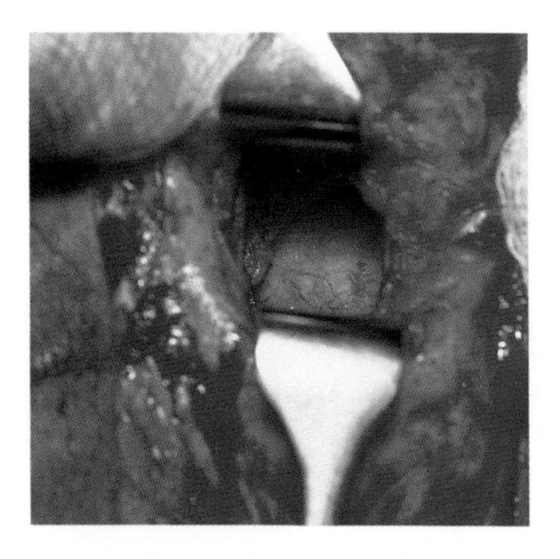

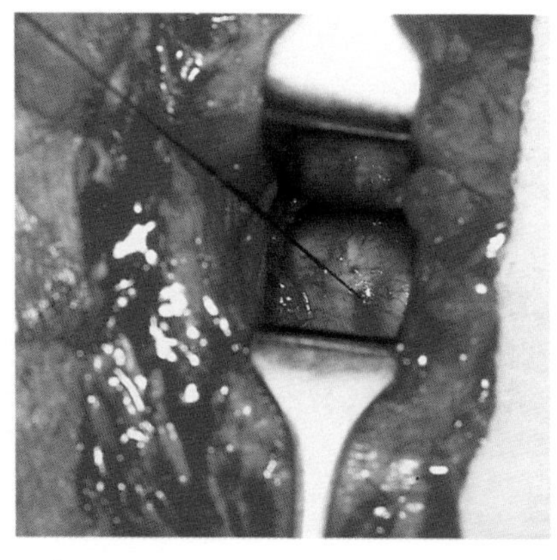

Fig. 4.14 Exposure of the trachea. (a) The trachea is exposed and its anterior wall cleaned by blunt dissection. This is kept to a minimum and is not allowed to extend laterally. The tracheal rings are counted by palpation and the thyroid isthmus is reflected superiorly.

(b) A strong stitch is inserted between the second and third tracheal rings; the needle is left in place.

(c) The stitch is held in artery forceps and used as a retractor.

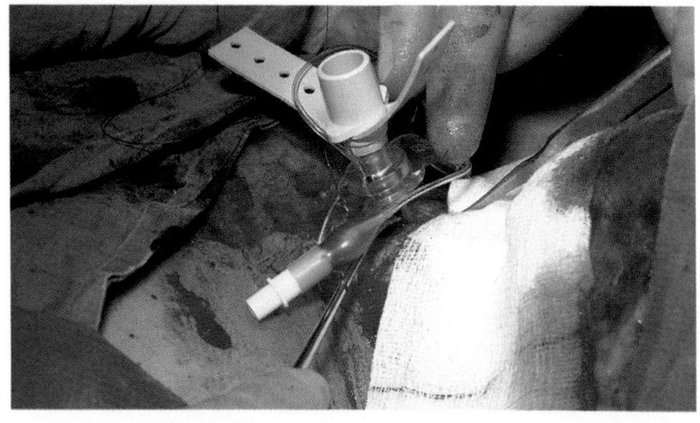

Fig. 4.15 Withdrawal of the endotracheal tube. An inverted U incision is made through the anterior wall of the trachea using a scalpel. The retracting stitch allows the flap to be pulled downwards. The endotracheal tube is revealed beneath, and the cuff should be deflated. The tube is withdrawn slowly.

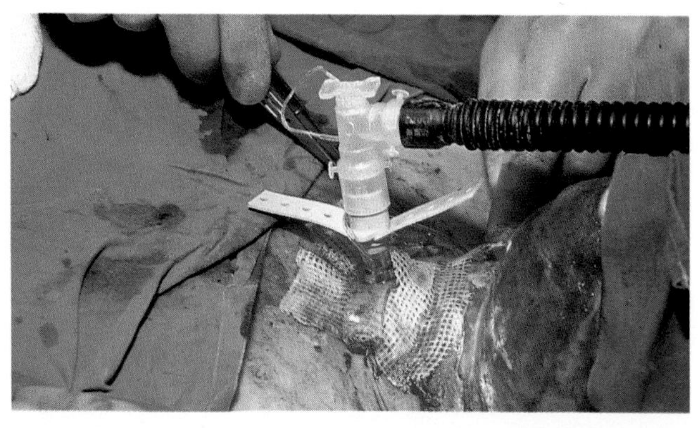

Fig. 4.16 Tube insertion. A large tracheostomy tube is inserted and connected via the adaptor to the anaesthetic circuit, and its cuff is inflated until an air seal is obtained.

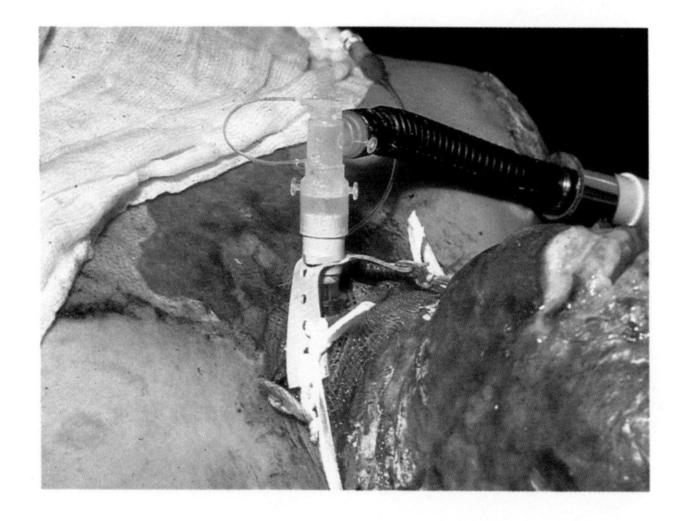

Fig. 4.17 Skin grafting. The graft is applied to the raw area and dressed. The strong stitch at the tip of the Bjork flap is passed through the graft and subcutaneous tissue at the inferior edge of the stoma. The flange of the tracheostomy tube is stitched to surrounding skin and held in place with the customary two tapes passed around the neck for extra security.

BURN WOUND INFECTION

A distinction must be made, both clinically and practically, between burn wound colonization and burn wound infection. A large, open area of damaged tissue, rich in nutrients, is very susceptible to contamination by bacteria from the patient's skin or secretions, other contaminated surfaces or the environment. The surface conditions provide an excellent nutrition medium for bacteria and will permit rapid logarithmic growth. As the eschar on the surface of the wound is isolated from the body, bacterial multiplication will not be inhibited by the normal defence mechanisms of antibodies and phagocytes. Contamination therefore leads to rapid colonization.

Burn wound infection occurs when the number of virulent organisms is sufficiently large to invade healthy tissues. Local resistance may then fail, with invasive sepsis leading to septicaemia and systemic infection.

In larger burns there is an element of immunosuppression due to a temporary failure in local cellular defences, with reduced chemotaxis, phagocytosis and clearance by the reticuloendothelial system, and by impairment in the specific responses against antigens. The loss of immunoglobulins is transient but T lymphocyte and macrophage activity is further reduced by immunosuppressive substances released by bacteria and damaged tissues. Under these conditions pathogens such as *Staphylococcus aureus* (Fig. 4.18) and the haemolytic streptococci (Fig. 4.19), together with the opportunist organisms *Pseudomonas aeruginosa* (Fig. 4.20), *Acinetobacter* and faecal streptococci, will thrive. *Proteus* infections may be seen anywhere (Fig. 4.21).

It is common to find a mixed bacterial flora on the burn wound, and secondary infection which may follow the use of antibiotics often involves multiresistant organisms and fungi.

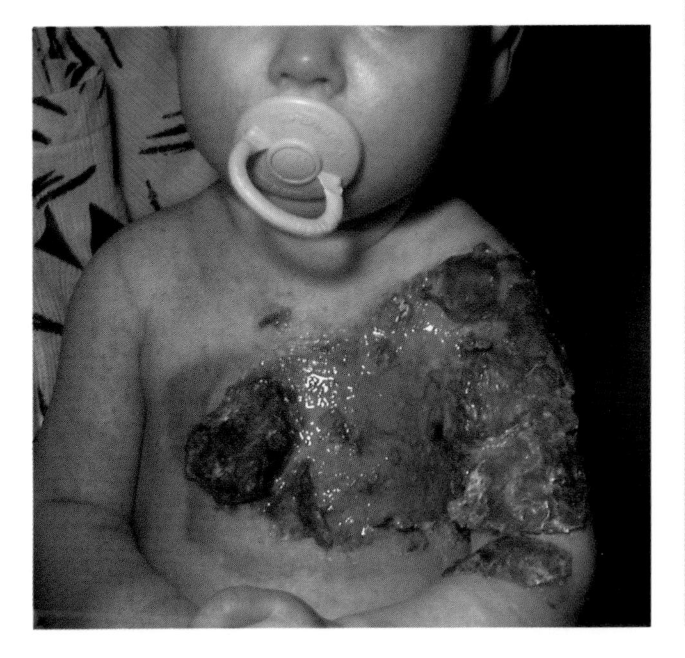

Fig. 4.18 Failed treatment by exposure method. The surface has been kept wet by saliva and the wound is heavily contaminated with *Staphylococcus aureus*. Thorough cleaning and dressing, with the administration of an appropriate antibiotic, are required.

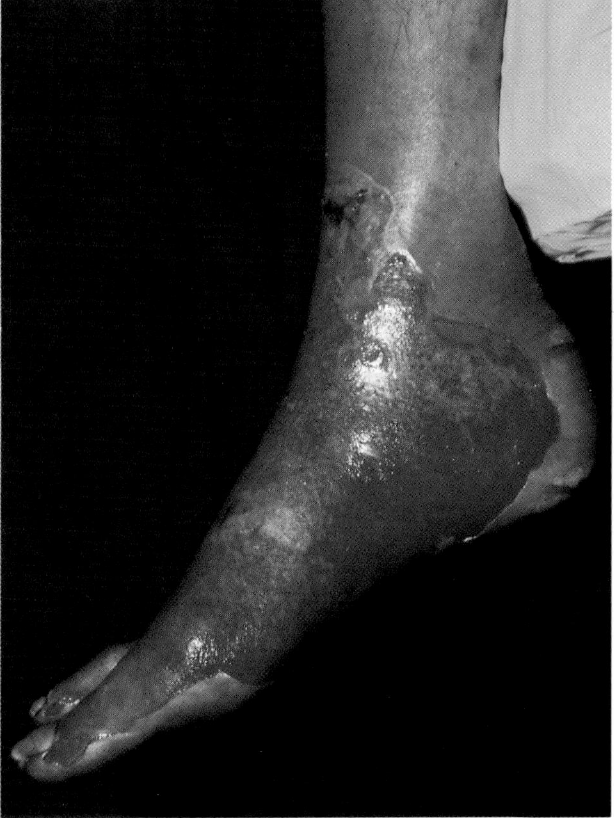

Fig. 4.19 Lancefield group A β haemolytic streptococcal infection of the foot. This is red, painful, rapidly spreading and accompanied by a marked rise in pulse rate, temperature and white cell count. The infection is very contagious, often seasonal and the organism is highly sensitive to penicillin.

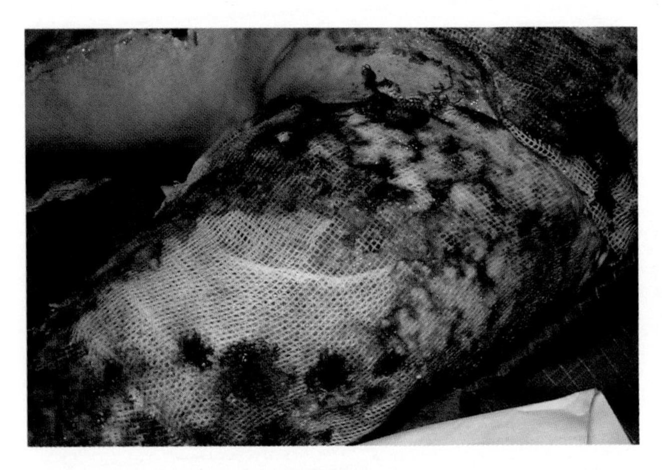

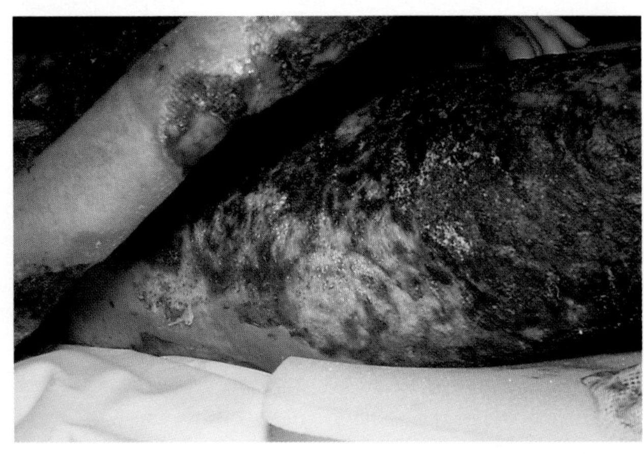

Fig. 4.20 Pseudomonal infections. (a) The green discoloration is typical. Burns around the groin and perineum are very frequently contaminated with *Pseudomonas*.

(b) Critical invasive infection due to *Pseudomonas aeruginosa*. The black lesions (ecthema gangrenosum) represent areas of vascular damage and infarction seen in burned patients in whom the normal immunological defence mechanisms are profoundly depressed.

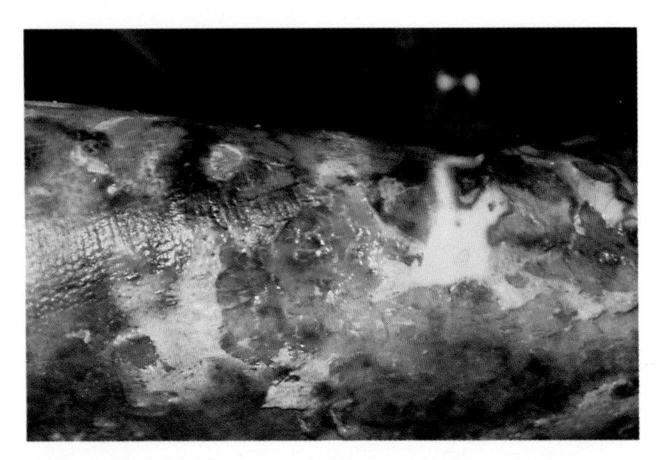

Fig. 4.21 Grossly infected burn wound. (a) The fat is invaded with Gram negative bacteria; the blue colour is associated with *Proteus* infections. Excision is required.

Prevention of contamination and colonization

Avoidance of contamination

Aseptic measures should be taken at all stages. The surface of the burn wound is initially sterile and every effort should be made to keep it in this state. The first aid covering should be sterile and easy to remove, and the wound should be inspected in a clean environment by personnel wearing masks, gloves, caps and gowns. Loose tissue is removed, the burn cleaned with an antiseptic solution, and a decision made on the method of treatment.

Patients with burn injuries must be nursed in an environment which minimizes exposure to extraneous organisms (see page 29). Isolated in a single cubicle, air is introduced under positive pressure so that it flows away from the burn and is expelled, allowing a continuous change of air. Attending staff should again wear protective clothing and ensure that hands do not come into contact with the burn wound. Meticulous attention is paid to washing the hands and arms to the elbows before and after coming into contact with the patient or the dressings. In this manner, there will be no cross-transference of organisms from the staff to the patient and vice versa. Most Gram negative infections are spread within a burn wound by contaminated baths or fingers. Gram positive organisms can be airborne. It is equally important for the attending staff to ensure that contamination from the patient's body fluids does not occur, thus eliminating the spread of viruses (particularly important in HIV-positive patients).

Dressings should be changed in a separate room which is well ventilated. Used dressings are removed and packaged, and gowns and gloves replaced before the new dressing is applied. Thorough washing and drying up to the elbows is important before the attendants leave a patient's room.

Exposure treatment
The burn wound surface can be made as inhospitable to bacterial growth as possible by encouraging it to dry to a hard eschar (Fig. 4.22). In a suitable environment where local humidity and temperature can be controlled, the exposure method is very effective. As the wound is open to inspection, any early signs of invasive infection such as cellulitis (Fig. 4.23) can be easily seen and quickly treated, but the dryness of the wound must be maintained and the eschar must be lifted off after two weeks (Fig. 4.24). Once the surface has become moist and macerated, rapid colonization is inevitable and the method must be abandoned.

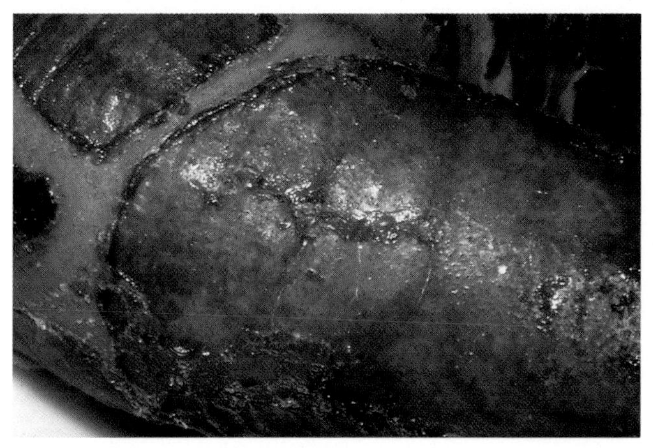

Fig. 4.22 Dry exposed eschar. This is an excellent barrier to infection, as the surface is relatively inhospitable. Gram negative organisms will dessicate and die.

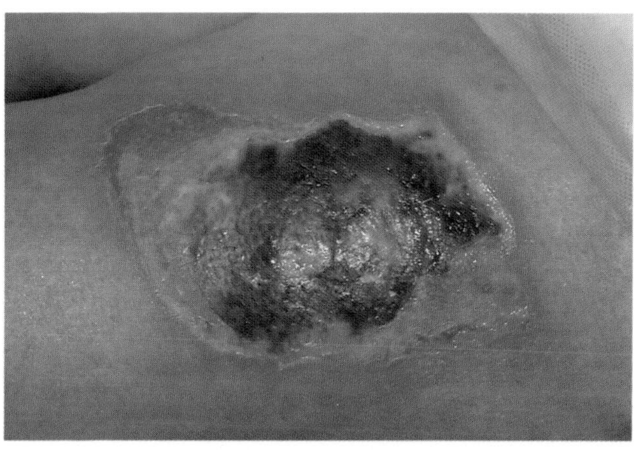

Fig. 4.23 Cellulitis and early invasive infection. This is indicated by the red flare around this small deep burn. Infection is likely to be due to staphylococcal or occasionally streptococcal (groups A, C or G) organisms.

Fig. 4.24 Eschar with partial infection. (a) Seemingly clean and dry eschar, excised from a full thickness burn, shows pockets of thick creamy staphylococcal pus.

(b) Undersurface of the same excised eschar; infection has been masked. It is unwise to leave a dry eschar on the burn surface for more than two weeks, as these small foci of infection become increasingly troublesome.

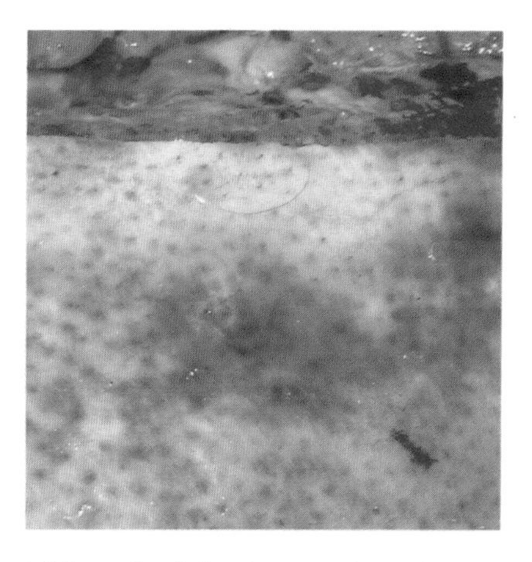

(c) Invasive infection at edge of escharotomy incision. The spreading brown stain on the skin is due to vascular damage and typically follows heavy colonization with *Pseudomonas*.

Topical antibacterial prophylaxis

An alternative method is to cover the wound with a dressing which acts as a barrier to contamination. Small burns may be dressed with a layer of vaseline-impregnated gauze and many layers of adsorbent covering, but in larger burns it is usual to include a topical antibacterial or antiseptic substance. Silver nitrate (0.5% solution) applied as a wet dressing is considered the best prophylaxis, but is expensive and time-consuming to apply, leads to staining, and can cause serious metabolic problems from the loss of chloride ions and methaemoglobinaemia.

Silver sulphadiazine cream can be incorporated in a dressing or 'buttered' onto the surface (Fig. 4.25). Sulphamylon (or mafenide) can penetrate the eschar, but is often painful to apply and can result in metabolic problems due to carbonic anhydrase inhibition. Povidone iodine, chlorhexidine and nitrofurazone, amongst others, are also used and can produce good results in the swollen burn. Nitrofurazone (furacin) is a very useful agent with a good spectrum against most organisms and the added advantage of drying the wound with the development of a neo-eschar, which forms as the necrotic tissue degenerates.

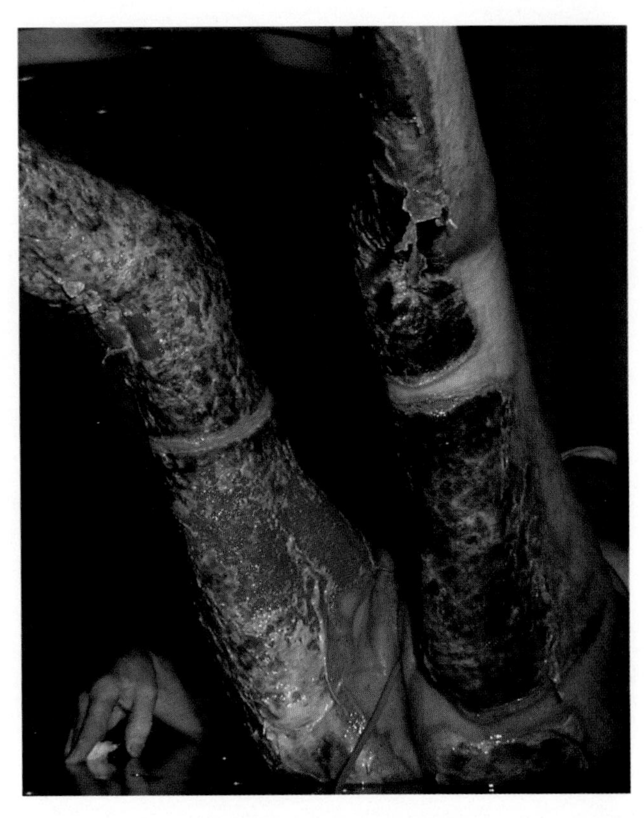

Fig. 4.25 Silver sulphadiazine dressings. This causes black staining which accumulates in the eschar. Slow separation reveals healthy granulation tissue underneath. There is no evidence of invasive infection.

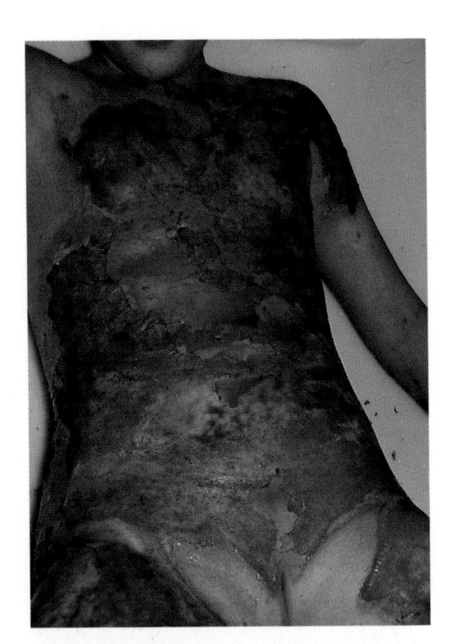

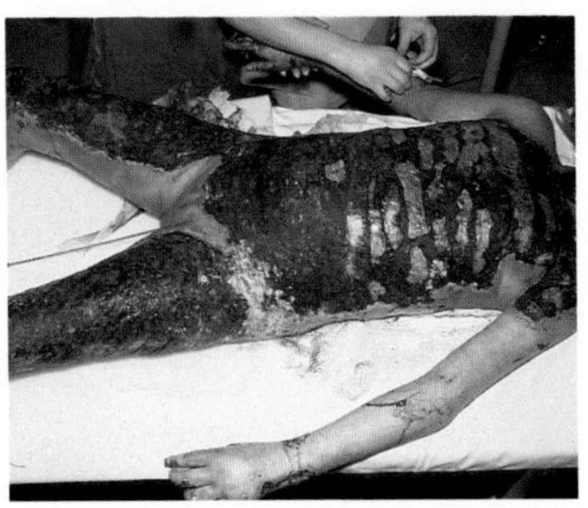

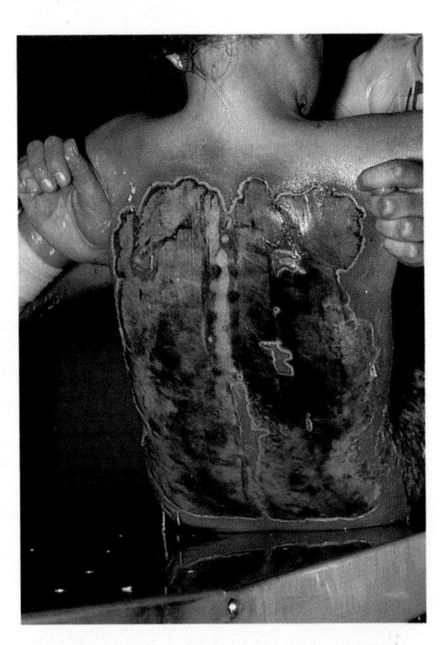

Fig. 4.26 Deep flame burns with complications.
(a) Ten days after the accident and seven days before surgery, the exposed wound appears clean and there is no bacterial growth from the wound swabs.

(b) Ten days after surgery, jaundice, *Pseudomonas* septicaemia and an obvious invasive burn wound infection have developed. There has been no response to gentamicin (note that the attendant nurse is not wearing gloves or a gown).

(c) The donor site on the back has been converted to a full thickness skin loss by the infection.

Prophylactic antibiotics
Although many authorities have recommended the use of prophylactic penicillin to prevent infection from *group A haemolytic streptococci*, its use is only beneficial in cases of local outbreaks of infection. In most other circumstances indiscriminate use of antibiotics encourages the emergence of resistant strains.

Early surgery
Early excision and grafting of the burn wound, before significant colonization has occurred, is advantageous but difficult to achieve. If infection does occur after surgery, treatment is directed towards antibiotics and intensive nutritional support (Fig. 4.34a). However, when early surgery is carried out systematically on all patients treated, thereby eliminating suppuration of wounds, it has a dramatic effect on the bacterial contamination of the environment.

Prevention of invasion and infection

Surveillance of bacterial cultures
Regular monitoring of the sensitivity patterns of organisms growing on the wound surface enables informed decisions to be made when the use of antibiotics is necessary (Fig. 4.27). Bacteria can be grown

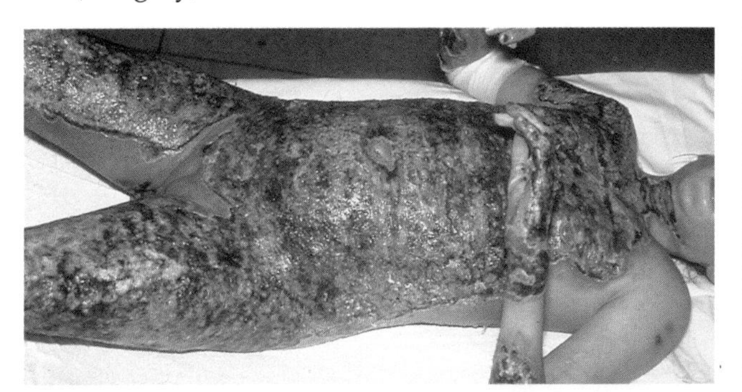

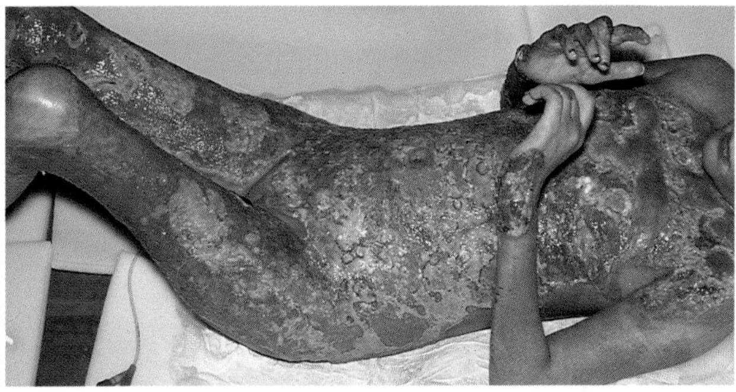

(d) Two separate septicaemias have been overcome with ceftazidime and intensive nutritional support, and the necrotic tissue is separating.
(e) Eleven weeks after injury, much of the dead tissue has been removed and some areas have been grafted with meshed skin.

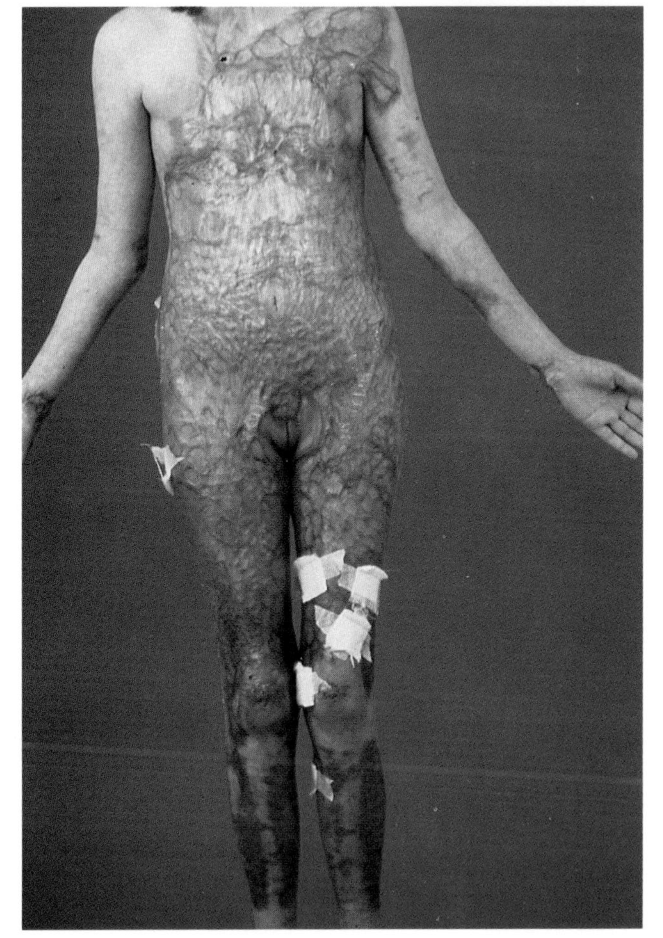

(f) Healed at 21 weeks.

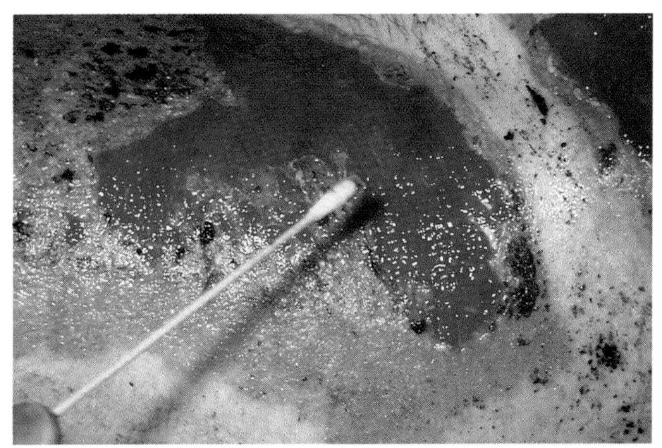

Fig. 4.27 Surveillance of bacteria on the burn surface. Monitoring should be carried out at least twice weekly. Regular assessments of antibiotic sensitivity patterns will enable early treatment with the appropriate antibiotic if a serious infection should arise.

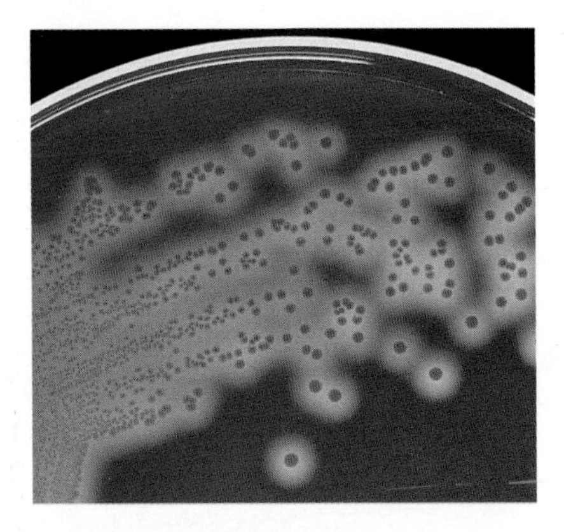

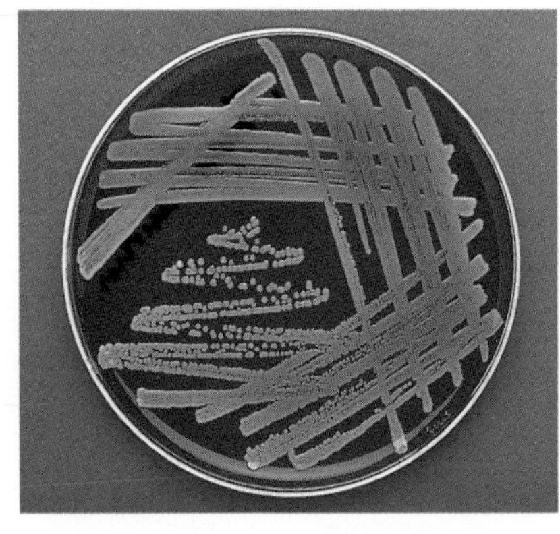

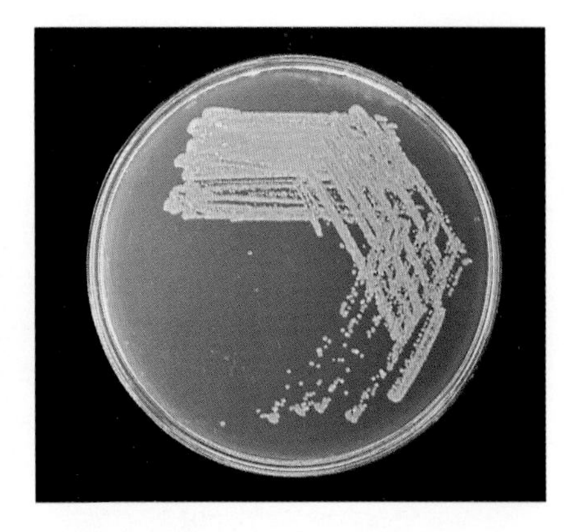

Fig. 4.28 Surveillance of bacteria using spread plates. (a) Appearance of Lancefield group A β haemolytic streptococci. Growth on a blood agar plate, with resultant haemolysis.

(b) Appearance of *Staphylococcus aureus* colonies. The medium is blood agar.

(c) *Pseudomonas aeruginosa* growth on King's media. This enhances the production of pyocyanin which is formed as a by product in the metabolism of *Pseudomonas* and has an antibiotic activity against Gram positive bacteria.

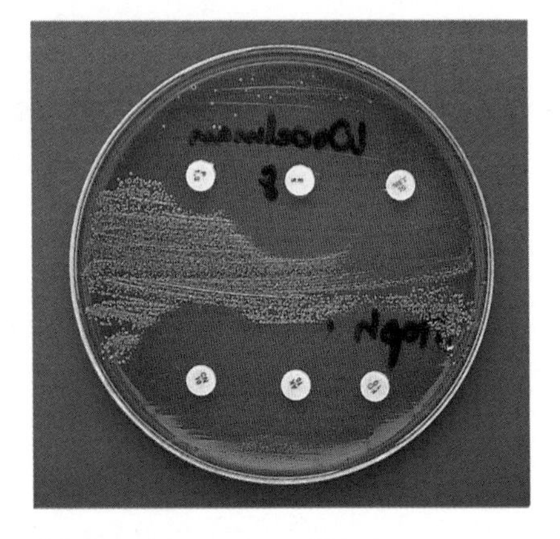

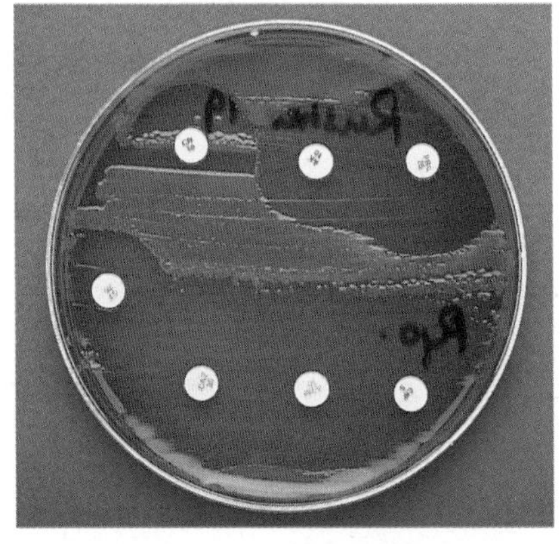

(d) Sensitivity plate. This is demonstrating *Staphylococcus* resistance to penicillin.

(e) *Pseudomonas* sensitivity plate demonstrating resistance to gentamicin.

on the relevant media and the resistance to various antibiotics confirmed (Fig. 4.28). Quantitative evaluations using wound biopsies may be helpful, particularly when invasive infection is suspected. The depth of such invasion can be seen with the use of specific staining techniques. This can be of great value in fungal and Gram negative infections.

Nutrition

Maintenance of adequate nutrition is of vital importance in resisting and combating infection. In the burn patient, the massive energy demands, together with cellular and protein losses must be compensated for before a significant deficit occurs. This may be achieved with regular blood transfusions and a very high protein diet via nasogastric intubation.

Fig. 4.29 A grossly infected burn wound. Little can be done to change the flora until all necrotic tissue is removed by excision.

Treatment of established infection

In most larger burns significant infection will arise at some stage (Fig. 4.29). None of the dressings available will sterilize a wound, and at best they manage to hold the numbers down to controllable levels. An infected wound must be dressed frequently, and it is the cleaning of the wound rather than the application of an antibacterial dressing that is important.

In the elderly patient with gross infections (Fig. 4.30), age and general frailty precludes any radical approach, and measures should be directed towards daily baths with debridement, systemic antibiotics and supportive measures of a high protein diet, maintaining mobility and blood transfusions.

Systemic antibiotics

Knowledge of the sensitivities of the bacteria growing on the wound will enable a sensible choice of antibiotics to be made. Several organisms are usually present and the most clinically significant should be identified. Attempting to cover all possible causes of infection with multiple or broad spectrum antibiotics encourages the development of resistant strains; an alternative is to eradicate the organisms likely to be responsible with a selective, narrow spectrum antibiotic.

The choice of antibiotics is extensive but consideration must be given to expense, toxicity and method of administration (see Table 4.3). The cell walls of many Gram negative bacteria contain β-lactamases (enzymes which destroy antibiotics) and it is their presence which makes these organisms, in particular *Pseudomonas*, so resistant to treatment.

Although in vitro studies may show that an organism is sensitive to a specific antibiotic, it is often found in practice that the latter fails to eradicate the target organism. Many antibiotics diffuse poorly, failing to penetrate the wound and necrotic material, and allow the organism to persist and develop resistance to the relatively poor concentration to which they are exposed.

Topical antibacterials

Few topical antibacterials have any activity against incipient or established burn wound infection as they cannot penetrate the eschar or damaged tissues. Thus they are of limited use. Bactroban (Mupirocin) is very effective in treating early streptococcal and staphylococcal infection; nitrofurazone, by tending to dessicate rather than macerate the burn wound, is useful in the early stages. Topical antibiotics are rarely used because resistance develops very rapidly. Neomycin has been used in spray form in the treatment of scald injuries until it was discovered that it caused permanent deafness in young children. Caution must be exercised in applying any aminoglycoside to a large burn wound. It must be remembered that the burn wound is absorbent and anything applied can be absorbed through it.

Surgical excision

Every attempt should be made to remove heavily infected and necrotic tissue when this can be done without exposing the patient to further risk. Under broad spectrum bactericidal cover at high dose, and with full supportive measures to replace blood, plasma and volume loss rapidly, the seriously infected burn wound may be excised down to fascia and immediately covered with viable skin. This is an extreme measure, but may be life saving; teamwork, timing and an ongoing commitment to repeat operations, anaesthesia and dressings are essential. More often, heavily infected tissue is removed without exposing further tissue planes to contamination. Daily dressings and wound cleaning are carried out under general anaesthesia and skin grafting delayed until the underlying bed is clean. If the graft is applied to a heavily colonized bed (Fig. 4.31), the skin graft will be lost and a repeat treatment will be necessary.

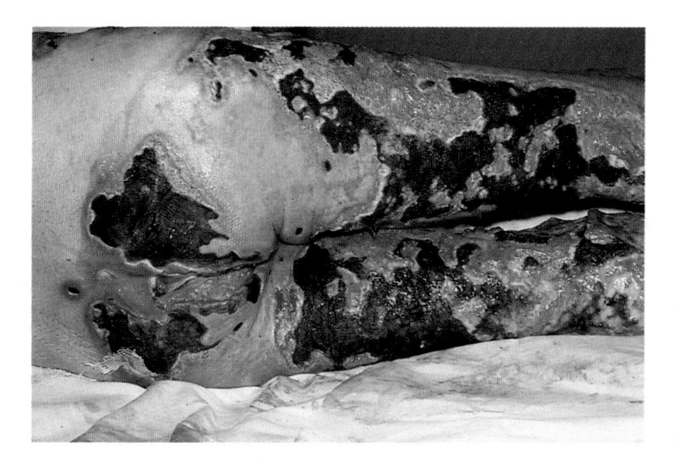

Fig. 4.30 Gross infection in deep burns by *Streptococcus faecalis* and *Pseudomonas aeruginosa* in an elderly patient.

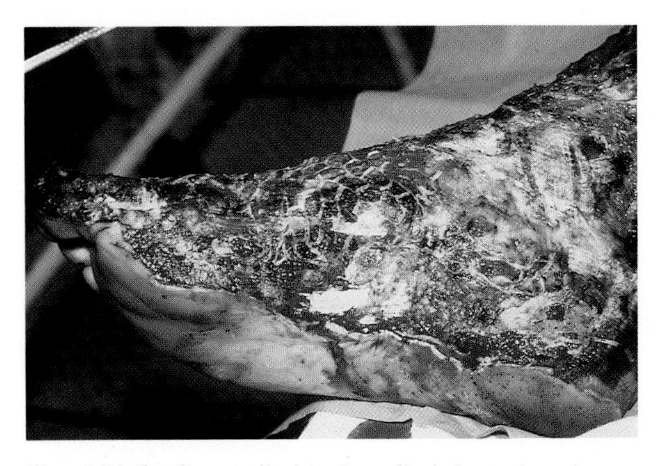

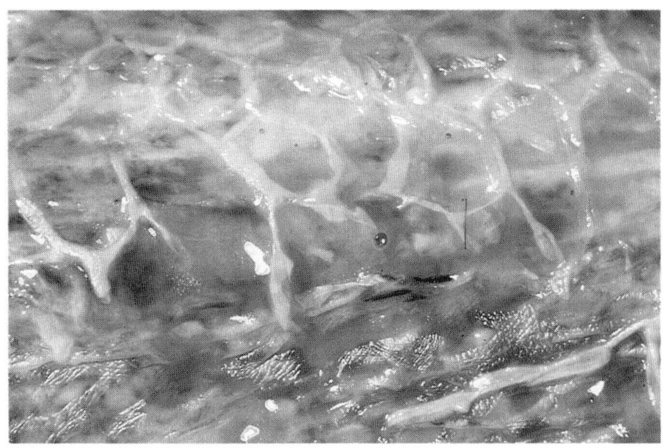

Fig. 4.31 Grafts applied to heavily infected surfaces. (a) Contamination is by *Pseudomonas* (b) Contamination by staphylococcal organisms.

Table 4.3 Systemic antibiotics

Antibiotic	Sensitivity	Comments
Aminoglycosides	Effective, broad spectrum	Toxic - serum levels must be monitored
Penicillins	Treatment of Gram positive organisms Effective against Lancefield group A β haemolytic streptococcus	Safe except where allergy persists
Flucloxacillin	For virulent staphylococcal infections	
Ticarcillin Piperacillin Azlocillin	Useful activity against Gram negative organisms	
Ceftazimide Imipenem Ciprofloxacin	Activity against Gram negative organisms, particularly *Pseudomonas* and *Acinetobacter*	No serious side effects

Septicaemia

Septicaemia resulting from invasive infection may present insidiously or in a florid manner. Typically, Gram negative sepsis is of slow onset with the temperature, white cell and platelet counts falling to subnormal levels. There is often abdominal distension, development of paralytic ileus, a fall in urinary output and delirium, although blood cultures are frequently negative. A rapid rise in temperature, with a swinging pyrexia, raised white cell count and rather severe confusion are characteristic of Gram positive septicaemia, but again blood cultures may be negative. Treatment in both cases is directed towards physiological support, i.e. rapid transfusion of blood to correct anaemia, plasma volume expanders to improve tissue perfusion, and oxygen. High doses of appropriate antibiotics are given intravenously and reviewed regularly. It is advisable to replace all intravenous lines on suspicion of septicaemia. Cannulae are rapidly colonized and the sites of previous cutdowns should be inspected for signs of cellulitis or thrombophlebitis. Mortality from septicaemia in burns remains very high, and the risk of septicaemic episodes will persist whilst necrotic tissue, immunosuppression, malnutrition and intravenous cannulation are present.

TRAUMA: METABOLIC CONSEQUENCES OF INJURY

A burn injury may have a profound effect on the metabolism of the body. There is an immediate change from the normal state of anabolic storage to one of catabolic breakdown as energy expenditure rapidly increases, which in turn increases nutritional requirements. Stress from anxiety, pain, heat loss, dressing changes, anaesthetics, surgery and infection also dramatically increases energy requirements. The basal metabolic rate may have doubled in a patient with burns of greater than 50 per cent of the body surface area; this state of hypermetabolism will persist until the wounds are almost healed. In order to maintain the huge and long-lasting demand for energy, the body draws on all available stored sources, including muscle. Advances in burn care have led to a better understanding and management of these energy requirements, which promotes a greater chance of survival, increases resistance to infection, and enhances rapid healing of donor sites and better graft take.

Metabolic considerations

During trauma, the instant secretion of catecholamines (adrenaline and noradrenaline) mobilizes glucose reserves which are released into the circulation and are rapidly consumed by the tissues, providing energy. This is followed by an increase in the levels of insulin, glucagon, growth hormone and glucocorticoids.

In the burned patient, glucose is mainly derived from muscle breakdown which is then selectively utilized in the damaged tissues. Glucagon stimulates hepatic gluconeogenesis, which utilizes substrates derived from the proteolysis of skeletal muscle. Liberated nitrogen is converted to urea and excreted. The urinary nitrogen level is therefore a useful measure of protein breakdown; when added to the loss of protein from the wound, this will give an overall indication of total protein requirements. Lipids, although rich in energy, require carbohydrates for breakdown by catecholamines and glucagon; this process is disrupted and inefficient in the burned patient, therefore lipids play little part in the production of glucose.

Temperature regulation is also disturbed. Homeostatic mechanisms in the hypothalamus are altered, causing a rise in core body temperature. The insulating effect of skin is lost through a burn injury which results in direct loss of body heat. Further heat loss is incurred by evaporation of water from the wound surface.

Table 4.4 The Curreri and Sutherland estimations of caloric requirements

	Adults	*Children*
Curreri	25 kcal/kg BW +40 kcal/% BSAB (when > 20%)	60 kcal/kg BW +35 kcal/% BSAB*
Sutherland		
Calories	20 kcal/kg BW +70 kcal/% BSAB	40 kcal/kg
Protein	1g/kg BW +3g/% BSAB	2g/kg

*Under eight years of age
BW = body weight
BSAB = body surface area burned

Estimating caloric requirements

A high protein, high calorie diet is recommended. There are two formulae which are widely used in the estimation of caloric requirements: that of Curreri, popular in the United States, which possibly overestimates requirements; and that of Sutherland, which is popular in the UK (see Table 4.4). The ratio of non-protein calories to nitrogen that should be administered is still under debate: the current estimate is 100:1 up to 125:1. It is important that the patient receives all carbohydrates and proteins required, and that progress is monitored. The best method of nutrition is by mouth, and this should be encouraged where adequate quantities of food can be tolerated. Ordinary food may be offered, and the patient is encouraged to drink large amounts of milk. The 35 egg-a-day diet has been used to improve hyperalimentation. If nutrition is inadequate, gross emaciation, muscle wasting with deformity and joint stiffening can occur (Fig. 4.32). Wound healing is also delayed (Fig. 4.33).

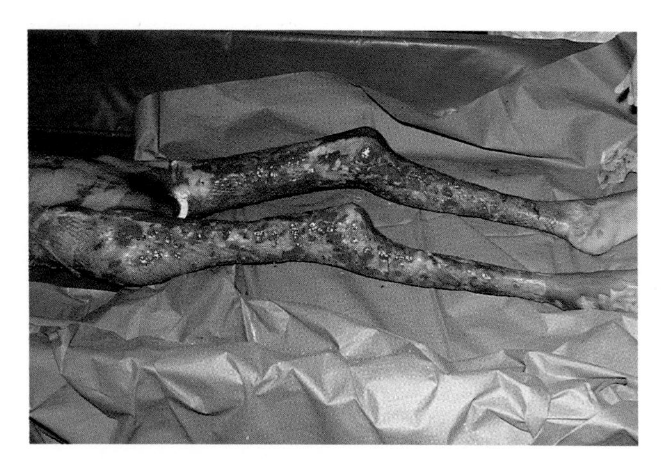

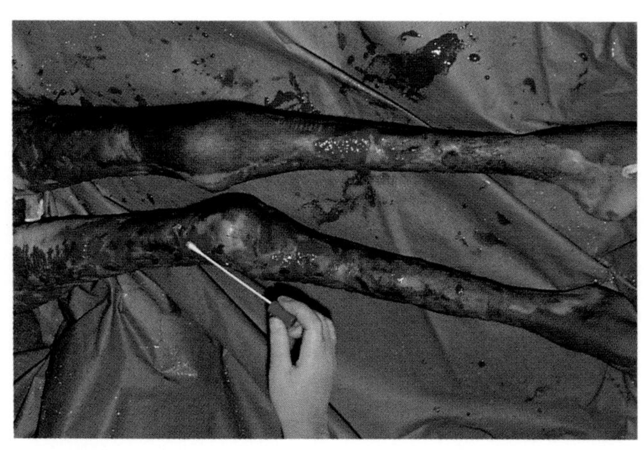

Fig. 4.32 Inadequate nutrition in a patient with extensive burns. (a) One year after injury. There is gross emaciation with extensive areas of unhealed burn. Almost total muscle wasting with considerable deformity and stiffening of the joints are seen.

(b) Over-granulating wounds with no development of epithelialization at the edges. As the wounds are small, spontaneous healing can be expected if nutrition is corrected.

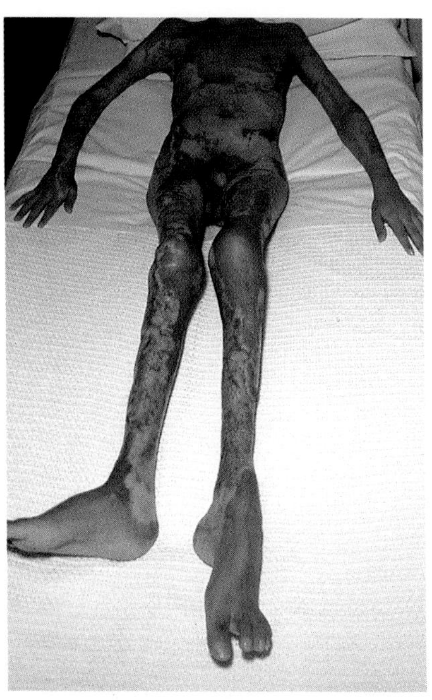

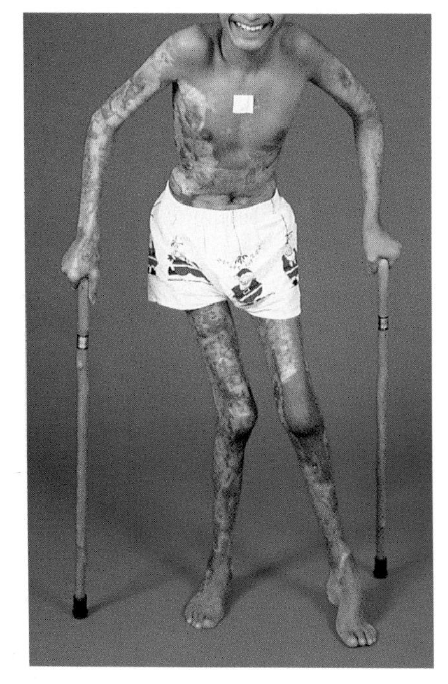

(c) Three months after continuous feeding through a fine bore nasogastric tube. All the burns have healed without the need for further grafting. There has been no appreciable gain in weight.
(d) One month later, the patient is able to support his own weight, but there is obvious deformity at the hips, knees and ankles.

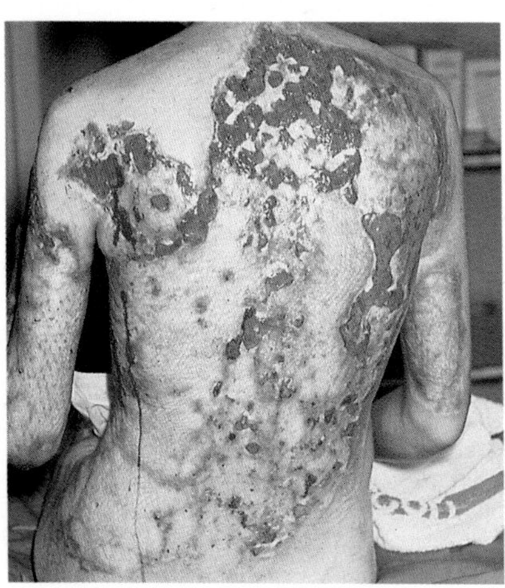

Fig. 4.33 Characteristic appearance of a chronic, non-healing burn wound. The pale, granulating areas are gradually becoming larger and the surrounding epithelium is white, macerated and non-adherent. The condition will be reversed if adequate protein is given.

Enteral feeding

Where it is not possible to administer sufficiently large quantities of food by mouth, which may often occur in cases with more extensive burns, nasogastric feeding is necessary. A fine bore nasogastric tube allows the administration of 2.5 to 3.0 kcal daily starting with half strength feed (50ml/h), the rate and concentration are gradually increased. Feed is continuously dripped into the patient via a simple pump-action delivery system. Formulae of high osmolarity should initially be used slowly and with caution. Nasogastric intubation is well tolerated, safe, and free from complications provided that the tube is correctly inserted in the stomach and not in the lungs (Fig. 4.34).

Parenteral nutrition

This method of nutrition allows a greater volume of fluid to be administered but it is only carried out in extreme cases such as periods of intense surgical activity, in those who are paralysed in order to assist artificial ventilation and cases of septicaemia. Great care and aseptic precautions should be taken with the insertion of intravenous lines. Central lines rapidly become colonized by bacteria, with frequent recurrence of bacteriaemias; these lines must be replaced frequently to minimize risk. Venous access may also cause difficulty as many access points may be burned. Parenteral formula must be high in protein, of low osmolarity and be carefully adjusted for carbohydrate content. A serious pseudo-diabetic non-ketotic hyperosmolar state may result from excess carbohydrate, with rapid dehydration from loss of glucose and water in the urine. Respiratory failure may also result from the high carbon dioxide load produced by glucose metabolism. Up to 25 per cent of the calorie requirement may be given as lipid.

Monitoring

Regular weighing of the patient is the most straightforward method of monitoring the adequacy of nutrition. Cases with burns of greater than 40 per cent of body surface area can lose up to 20 per cent body weight if metabolism is not vigorously supported. A weight loss of greater than 10 per cent is associated with a significant delay in healing and an increased incidence of sepsis. A weight loss of greater than 35 to 40 per cent, particularly if rapid, usually leads to death.

Biochemical estimations of the levels of plasma proteins and albumin are useful indications of the effectiveness of the metabolic response, but are slow to reflect change due to their long half life. Transferrin, having a higher turnover rate, is a more accurate indicator.

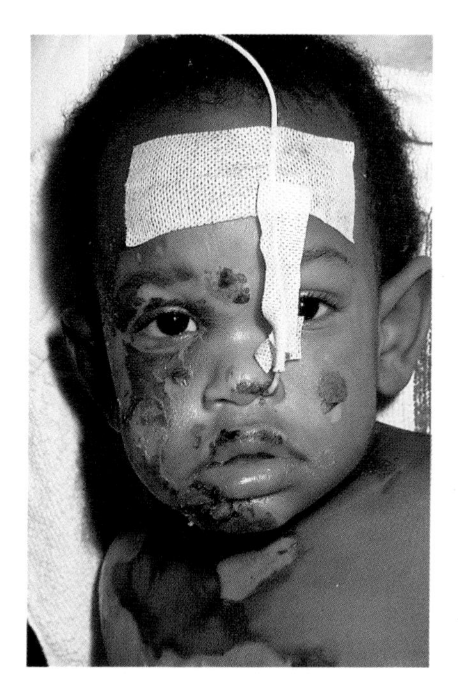

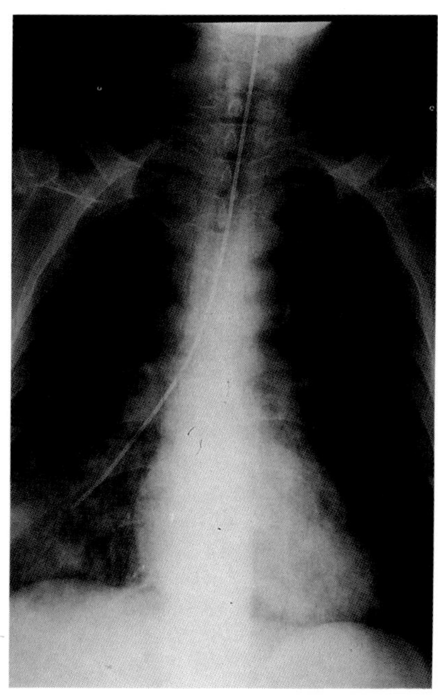

Fig. 4.34 Fine bore nasogastric tube. (a) This method is surprisingly well-tolerated in normal circumstances.
(b) X-ray of the chest demonstrating a fine bore tube inserted into the right main bronchus. A considerable amount of feed has gone into the lung. The tube had become displaced during anaesthesia and feeding recommenced before its position had been checked.

5 Late Management

ANAESTHESIA

Jonathan M. Chandy

Burns patients do not represent a typical cross section of the community. The extremes of age are over-represented, as are handicapped, epileptic, disadvantaged and psychiatrically disturbed patients.

Venous access

The sites normally used for intravenous infusions are frequently burnt. Moreover, even if intact, the veins will have been used already during the resuscitation phase. Neck lines can be inserted to cover the deslough, but in major burns this area is usually damaged. Subclavian lines should be X-rayed prior to the induction of anaesthesia. For a major excision and grafting, two drips should be set up. Establishing adequate intravenous access in small infants is problematic.

Airway problems

These are common in burns to the head and neck and form three distinct problem areas.

Early

Oedema of the face and upper airways is most severe during the first 72 hours. Maintaining airway patency can be difficult and intubation with a fibreoptic bronchoscope may be necessary in the awake state (Fig. 5.1).

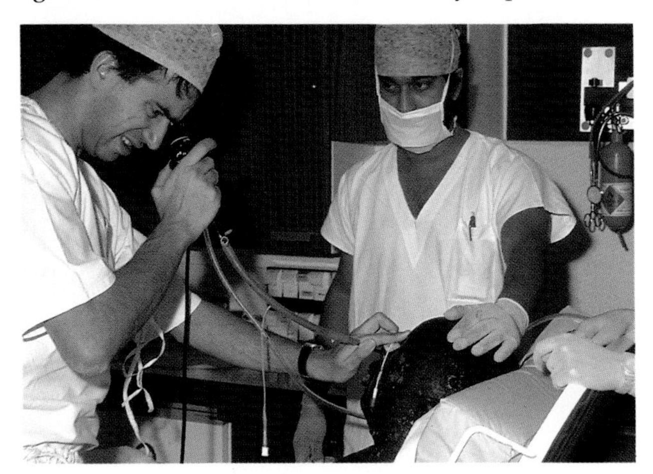

Fig. 5.1 Maintenence of airway patency. The endotracheal tube is mounted on the fibreoptic laryngoscope and passed through the larynx. The patient is under Ketamine anaesthesia.

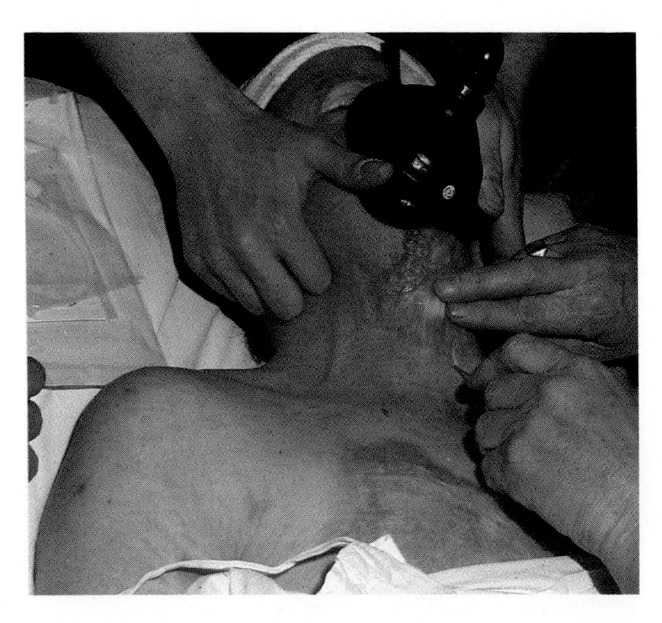

Fig. 5.2 Introduction of the endotracheal tube with restricted opening of the mouth. (a) Scar contracture between the jaws has made visualization of the cords impossible, and blind nasal intubation has been tried and abandoned. With the neck fully extended, a large gauge spinal needle is inserted via the cricothyroid membrane into the larynx.

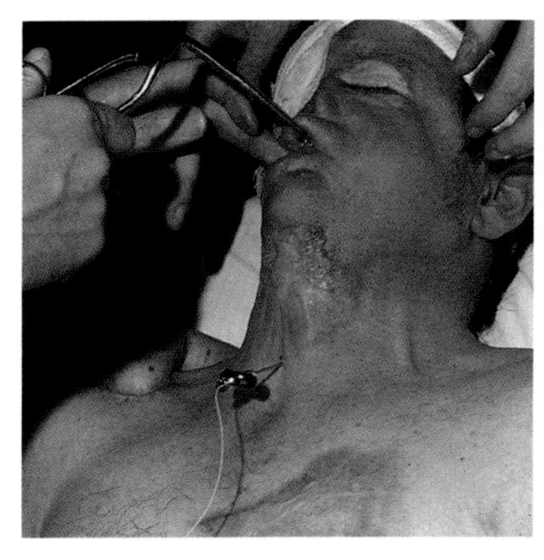

(b) A stiff polythene cannula has been passed through the needle into the larynx and retrogradely through the cords. It is retrieved from the pharynx with Magill's forceps.

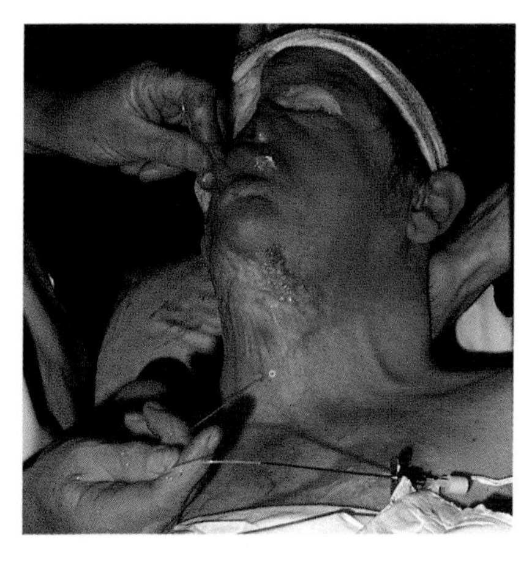

(c) The cannula is brought out through the mouth and an endotracheal tube passed down the mouth between the cords and into the trachea.

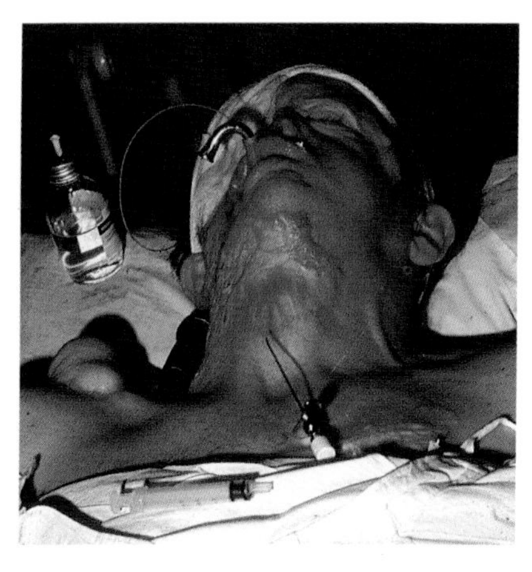

(d) The tube is in place and the cannula can be with-drawn. This method was first described by Waters.

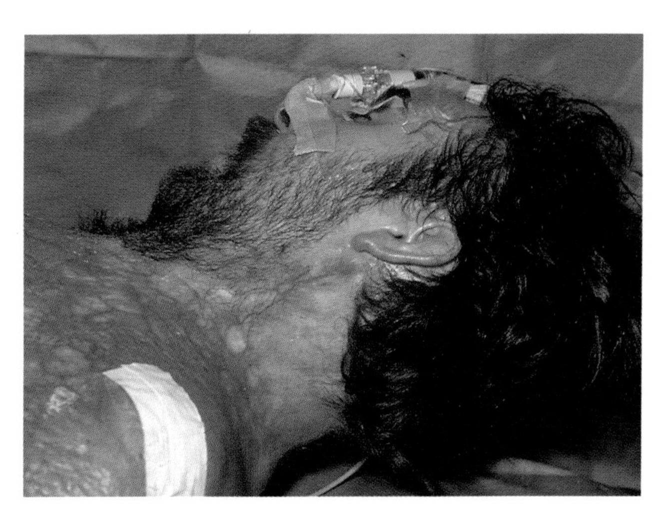

Fig. 5.3 Marked neck contracture. This makes intubation both difficult and dangerous.

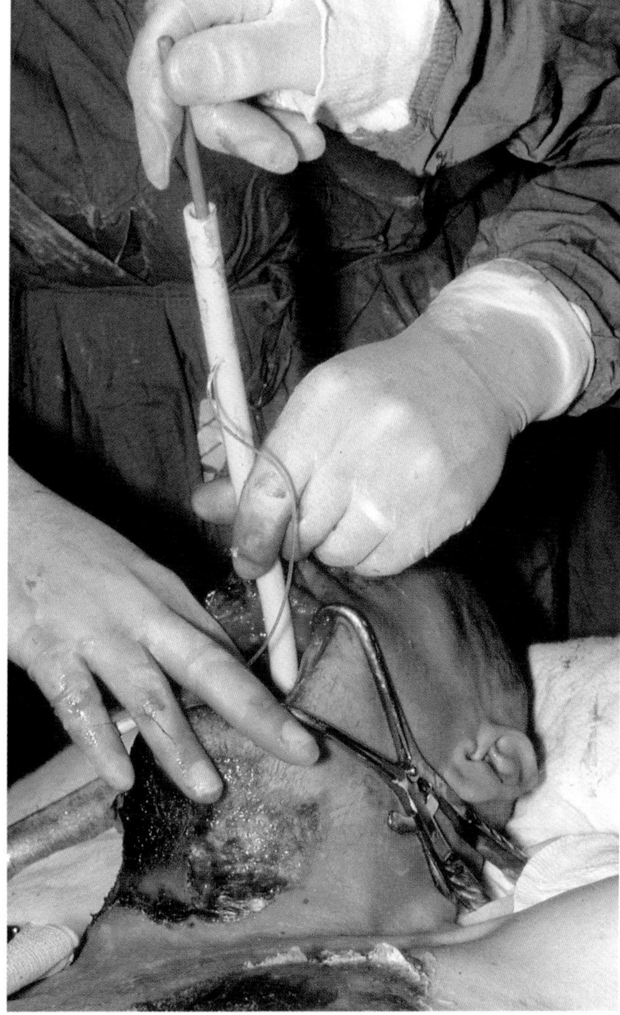

Fig. 5.4 Use of the Featherstone gag. The jaw has been held open with a Featherstone gag wedged against the molars, enabling examination of the airway with a laryngoscope by the anaesthetist.

Late
During late skin grafting and subsequent reconstructions, difficulties may arise where scarring has produced small, tight lips (Fig. 5.2) and fixed flexion of the neck (Fig. 5.3). In the former, the jaws may be held open with a Featherstone gag to allow the introduction of the endotracheal tube (Fig. 5.4). An additional complication could arise where the glottis has become tethered anteriorly to the tissues of the neck.

Inhalational injuries
There is usually considerable shunting following an inhalational injury. It is advisable to run the anaesthetic with a relatively high $_{Fi}O_2$, and to use a heat-moisture exchange.

Fluids

Measurement of fluid loss
Desloughing an extensive burn results in major blood loss. In adults, this is in the order of 100ml blood/% body surface area excised. Measurement of loss by conventional means is difficult, for the following reasons:

1. Bleeding occurs from the body surface rather than a body cavity, and cannot be collected by suction and measured.
2. The desloughed area is frequently dressed with hot saline soaks to reduce bleeding; these dressings contribute to the apparent loss.

Problems associated with massive transfusions
Patients with major burn injuries frequently undergo an exchange transfusion on the operating table, and are therefore subject to complications associated with massive blood transfusions. In order to minimize these, all blood should be warmed and fresh frozen plasma given routinely in cases where more than 30 per cent of the circulating blood volume has been replaced.

Postoperative fluids
The requirements for postoperative fluids are difficult to estimate. Although the patient may leave theatre in a haemodynamically stable position, bleeding can continue in the ward and may increase dramatically if the patient rewarms. It is advisable to prescribe enough fluid to cover the loss, and the fluid regime may then be adjusted by the houseman as appropriate.

Temperature

A fall in temperature may occur due to the effects of anaesthesia and from loss of skin due to the burn injury.

To counteract heat loss, the following measures may be implemented:

1. Maintenance of a high ambient temperature.
2. Heating coils for all fluids which are to be transfused.
3. A heated ripple blanket.
4. Heated humidification of inspired gases.
5. A prompt return of the patient to a heated post-operative environment.

Drugs

The patient with major burns may respond to drugs in a different way to the uninjured person. Some specific responses are listed below.

Suxamethonium
Administration of suxamethonium may result in a substantial increase in serum potassium. The flux is initially evident at about two weeks after injury, and can be seen for several months. The 'window of safety' for this drug is still under debate and seems to be narrowing. However, it is safer to administer suxamethonium rather than a non-polarizing muscle relaxant if there are doubts regarding the maintenance of airways or intubation.

Non-polarizing muscle relaxants
Patients with major burn injuries show resistance to non-polarizing muscle relaxants. This may be due to an increased volume of distribution, reflecting the hyperdynamic state of circulation, together with an increase in acetylcholine receptors. This latter phenomenon may explain the potassium flux seen with suxamethonium.

Enflurane
A significant number of burn victims suffer from epilepsy. Care should be taken to avoid using enflurane in these cases.

Halothane
Halothane hepatitis is extremely rare in patients with major burns; this may be related to the increased immunosuppression which is associated with this type of injury. In children, halothane hepatitis is relatively rare, but in children with burns it is almost unknown. It is therefore considered safe to administer repeat halothane anaesthetics. Moreover, establishing venous access in a conscious, post-resuscitation infant can be quite demanding, and isoflurane and enflurane inductions in small infants are not always trouble-free.

Opiates

(a) Operative

Patients usually display considerable tolerance to opiates. This may be due to the high degree of pain suffered and the fact that these patients will have been receiving maintenance doses for some time. Furthermore, a substantial proportion of intravenous narcotics administered during the anaesthetic will be lost through exsanguination in theatre.

(b) Postoperative

In major burns, opiate infusion is the most efficient method of delivering postoperative analgesia.

Calcium

Disordered calcium homeostasis is displayed in the burned patient. Precipitate falls in ionized calcium have been demonstrated which may explain the relatively high incidence of sudden death on the operating table. In order to prevent this, routine administration of 25–50mg calcium chloride (or 10mg for infants under 10kg) every 15–30 min has been suggested.

Induced hypotension

Formal, induced hypotension may appear to offer substantial savings in blood loss. However, the practicalities of protecting an arterial line in a patient who is rotated several times during the procedure, together with the considerable risk of losing control of the situation, render this method hazardous.

Monitoring

Blood pressure

(a) Direct

Measurement of direct arterial pressure can be advantageous in several ways. However, during anaesthesia, the difficulties of preventing a disconnection while the patient is continually turned can outweigh the advantages.

(b) Indirect

Non-invasive monitoring can normally be achieved at the start of surgery, but in the presence of burns to the upper half of the body, the cuff may have to be placed on the leg rather than the arm. However, if left in this position the cuff will often occlude the site of intravenous infusion.

Furthermore, as the legs are often used as donor sites for grafting, the cuff will need to be removed and blood pressure measurements will not be possible when the volume depletion is at its highest.

Capnometer

The measurement of expired carbon dioxide is possibly the most useful method of monitoring available and provides evidence that:

1. The ventilator is properly connected.
2. The minute volume is appropriate.
3. There is a cardiac output.

Pulse oximetry

This method of monitoring is very useful and should be employed whenever possible. However, burn injuries to the hands are frequently seen and therefore pulse oximetry may prove problematic.

Temperature

Core temperature is usually monitored using nasopharyngeal probes. If this falls below 34°C, surgery should be stopped as soon as possible and the patient electively ventilated postoperatively.

Electrocardiography

ECG monitoring may prove difficult during the deslough of a major burn. Standard electrodes (Fig. 5.5) tend to slip off during exsanguination once the surgery begins. Needle electrodes (Fig. 5.6) or crocodile clips attached to skin staples can ensure continued monitoring. Care should be taken not to use monopolar electrocautery when these unconventional ECG electrodes are used.

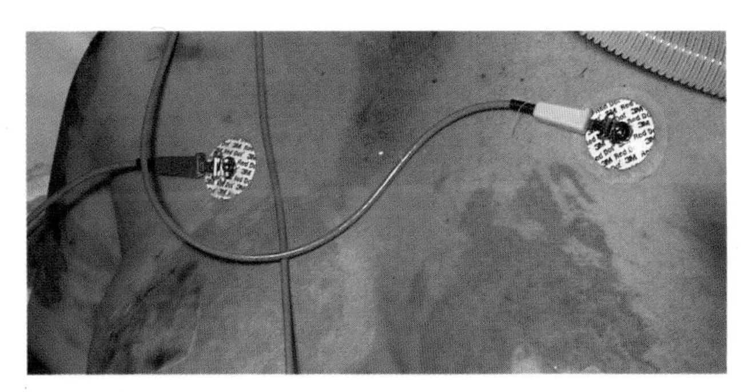

Fig. 5.5 Standard ECG electrodes. These are easily dislodged during operations on the burn-injured patient and are difficult to keep in place.

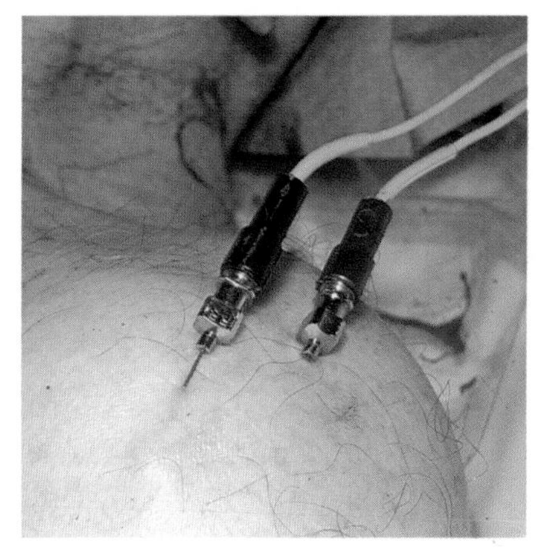

Fig. 5.6 Needle electrodes.
(a) Insertion into unburnt skin

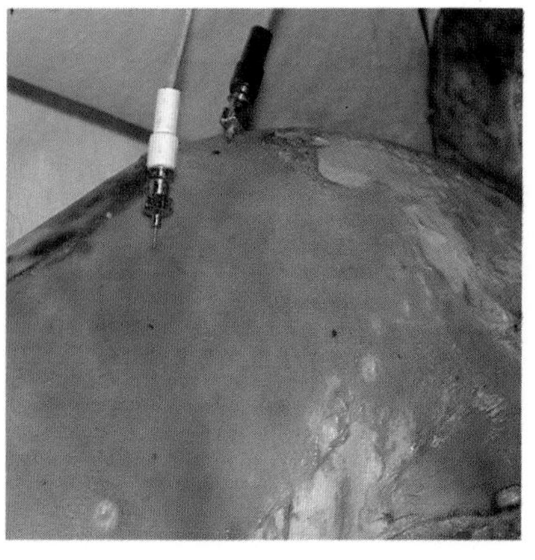

(b) These may also be inserted into burnt skin.

Fig. 5.7 Self-administered analgesia.

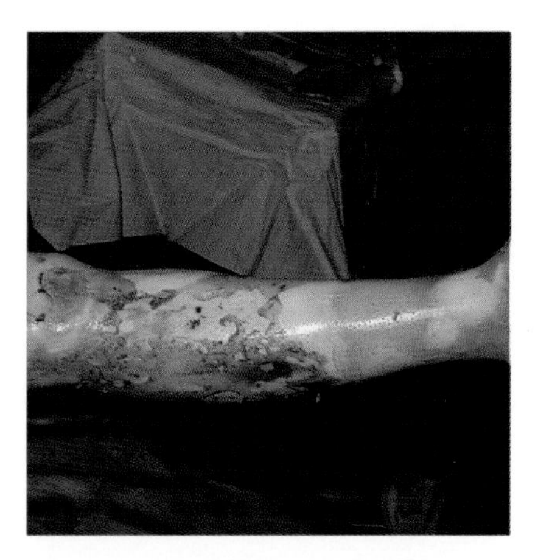

Fig. 5.8 Tangential excision.
(a) Deep flame burns of the leg, mainly full thickness in depth, but with areas of deep dermal injury.

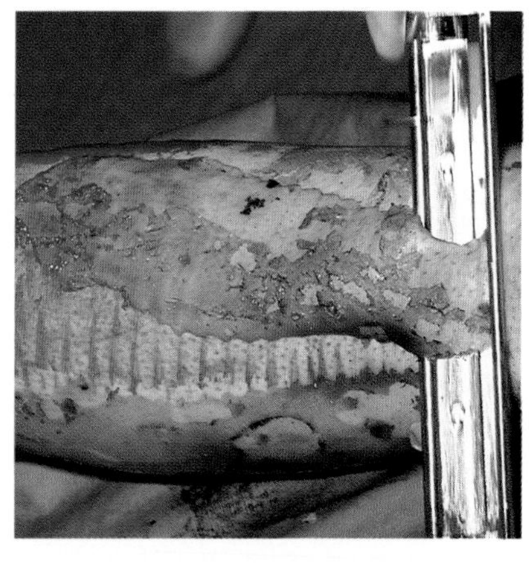

(b) Tangential slices are removed with a hand-held dermatome.

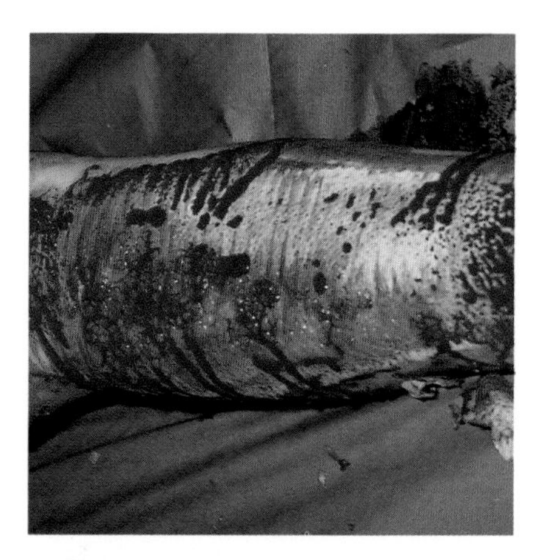

(c) A uniformly bleeding layer is reached.

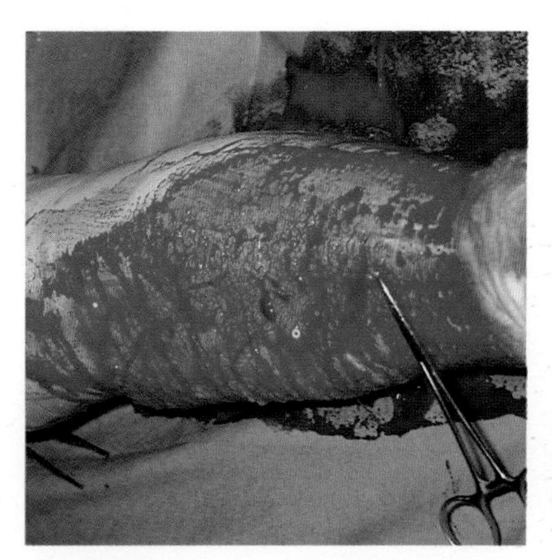

(d) Haemostasis is achieved with cautery, fine catgut sutures and the application of pressure.

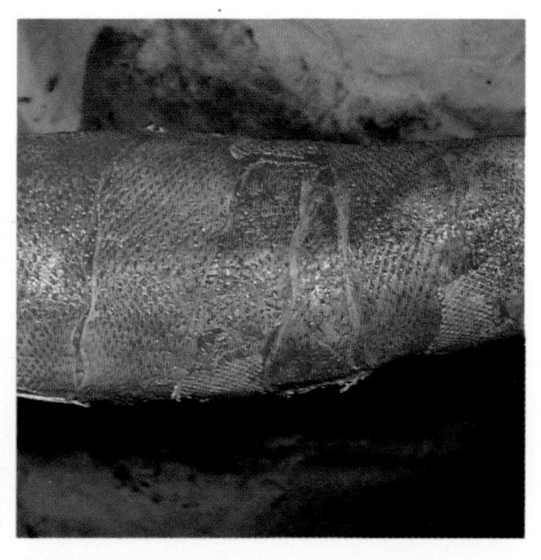

(e) The wound is covered with a finely expanded meshed skin graft.

Postoperative anaesthetic care

Feeding
The anaesthetic technique should be directed towards the earliest possible resumption of enteral feeding.

Intermittent positive pressure ventilation
Patients who have suffered a significant drop in core temperature, or are cardiovascularly unstable, should be considered for elective postoperative ventilation.

Self administered analgesia
Self administration of analgesia during baths and dressing changes can sometimes be encouraged (Fig. 5.7). Entonox (50 per cent oxygen, 50 per cent nitrogen dioxide) is excellent for this purpose. The patient must be confident in the technique of holding the mask on tightly and inhaling hard. Training is needed before the procedures are started.

Psychological considerations
Patients with major burns will often be embarking on a lengthy and painful course of surgery. Moreover, they will have to cope with the psychological traumas of scarring. Patience and reassurance by the anaesthetist from the start of treatment will prove highly beneficial in subsequent theatre episodes.

WOUND EXCISION

Ideally, wound excision should be carried out as soon as possible after injury as the wound is sterile and no infection has developed. The decision to delay excision is dependent on the availability of resources, personal preference and experience of the surgeon. In deep burns, where rapid spontaneous healing does not occur, excision should be performed early as infection may set in which will compromise graft take and increase blood loss at the time of surgery. Small, deep burns can be excised easily, with few complications. However, early excision of large burns is particularly challenging, as it presents extreme trauma to the patient and considerable support is required. Such surgery is major and should only be undertaken where facilities and commitment are available. Moreover, the excised wound must be closed with a skin graft or an effective biological dressing.

Tangential excision

The most widely used method for early excision was first described by Jancekovic and is mainly used in cases of deep dermal burns. The burn is shaved down to healthy bleeding tissue which helps to preserve the dermis and ensures good graft take, before any significant infection can develop (Fig. 5.8; see also page 68). Healing will then proceed rapidly. The risk incurred to the patient with extensive burns may be outweighed by the excellent results which can be obtained.

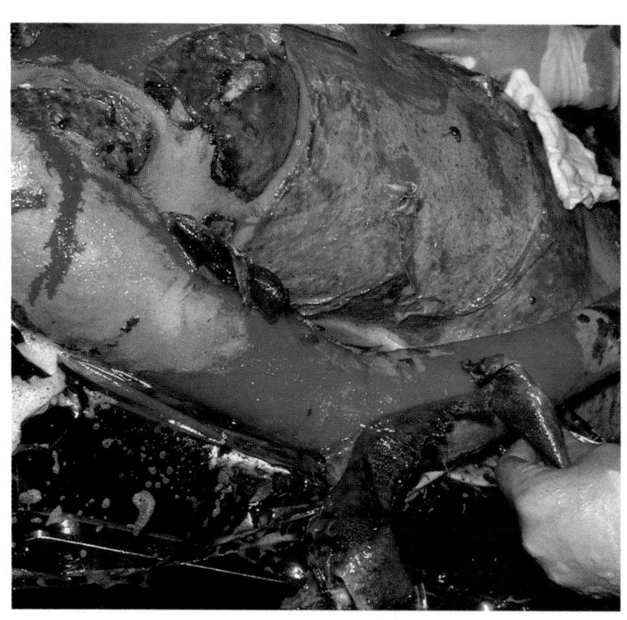

Fig. 5.9 Excision of an exposed burn. (a) After two weeks exposure, and under general anaesthetic, the hard, dry eschar is gently separated from the underlying tissues. A plane of separation can often be found with the fingertip, and the eschar is peeled off

(b) Blood loss is remarkably small and a skin graft can be applied to the wound bed without difficulty.

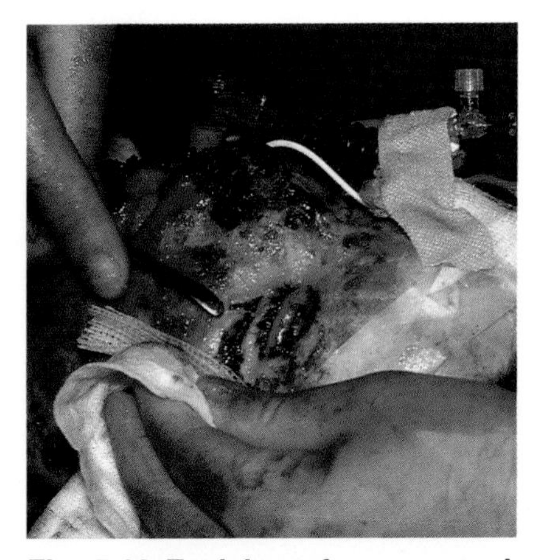

Fig. 5.10 Excision of an exposed burn using a skin graft knife.
(a) Adherent areas of eschar on the face are prised off.

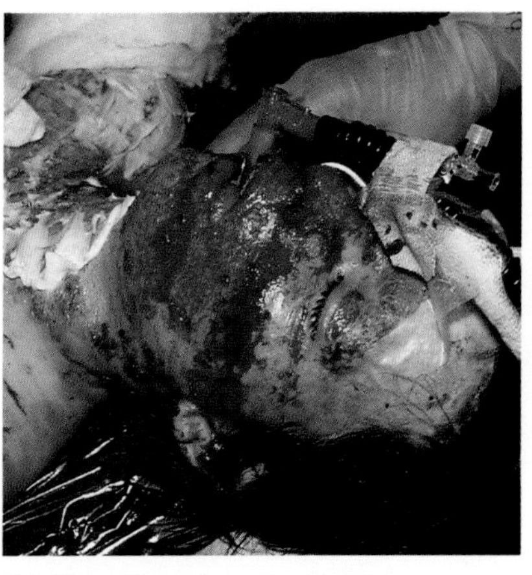

(b) Blood loss is only slight.

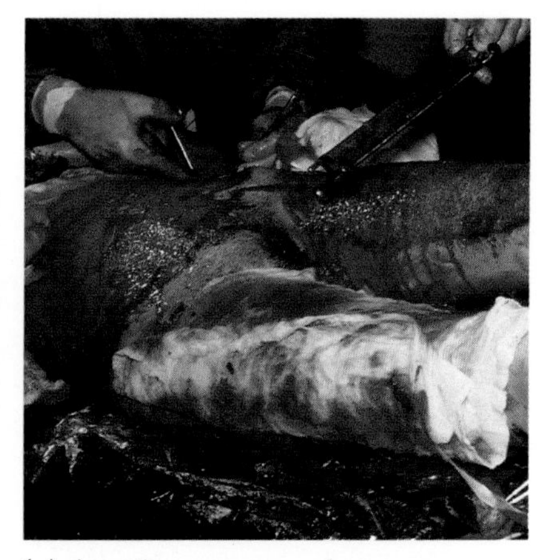

(c) In other areas, where no cleavage plane can be found, the eschar must be removed using a skin graft knife. Blood loss is more marked but is appreciably less compared with an infected wound, or after treatment in dressings.

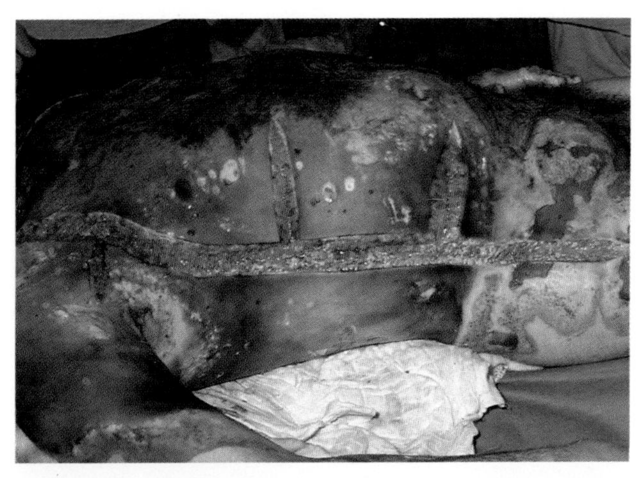

Fig. 5.11 Excision of a burn treated by the 'closed' method. (a) An old, infected deep burn with partial separation of the eschar.

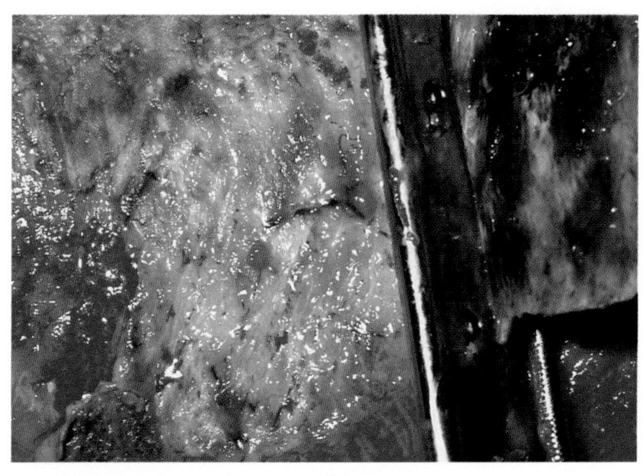

(b) There is obvious granulating tissue.

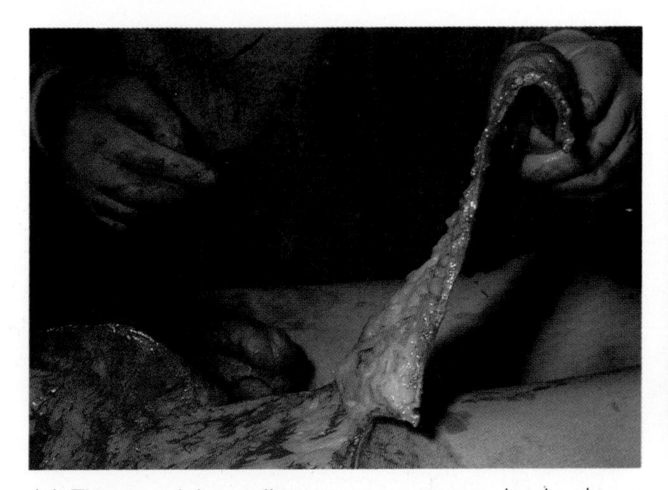

(c) The remaining adherent areas are excised using a skin graft knife and scalpel.

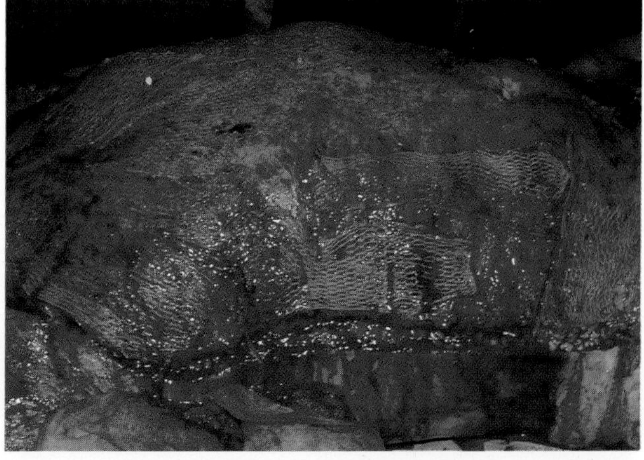

(d) The wound will be dressed until the bed is more suitable for grafting.

Excision of the exposed burn

Burns treated by exposure should be excised at two weeks after injury; subeschar sepsis and inflammation of the wound bed will be minimal, thus blood loss will not be excessive and uncontrollable. The eschar may be easily peeled off; a plane of separation can often be found with the fingertip (Fig. 5.9). Adherent areas of the eschar can be excised with scissors, but in areas where no cleavage plane can be found the eschar must be removed using a skin graft knife (Fig. 5.10).

Excision of burns treated by the 'closed' method

This type of burn shows a slow separation of the eschar with maceration and inflammatory changes in the wound bed. Excision is technically difficult and accompanied by severe blood loss. Haemostasis is difficult to achieve, which can compromise graft take, or necessitate delayed application of the skin graft (Fig. 5.11).

Excision to fascia

In cases of severe wound sepsis or in large deep burns, a case can be made for wide and deep excision of involved tissues (Fig. 5.12). In these cases, it may have been necessary to perform a fasciotomy in order to decompress swollen muscles and release the blood supply (Fig. 5.13). During excision a high level of support is needed, including appropriate antibiotic cover. The wound must then be closed with viable skin however. Excision that extends only down to a layer of healthy fat helps to improve a cosmetic appearance (Fig. 5.14).

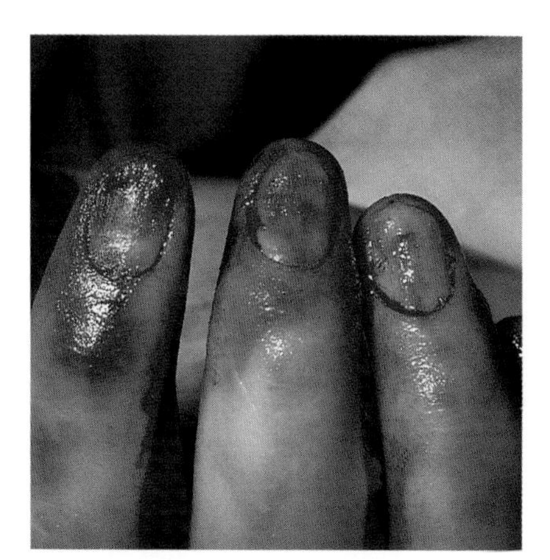

Fig. 5.12 Excision to fascia.
(a) Deep flame burns of the hand, with loss of the fingernails.

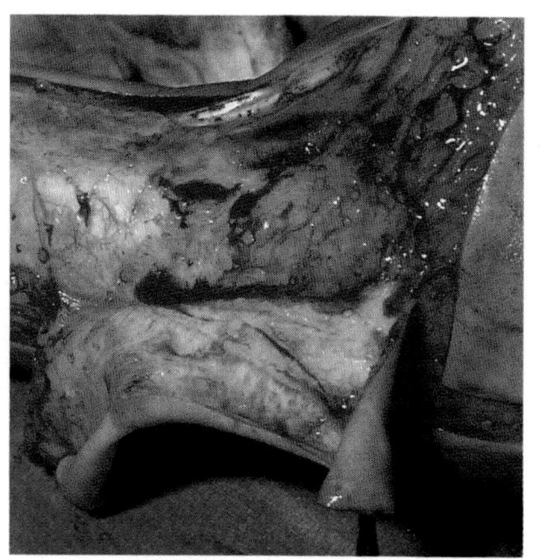

(b) An escharotomy was performed on admission. Twenty-four hours later, sharp excision to the fascia was carried out.

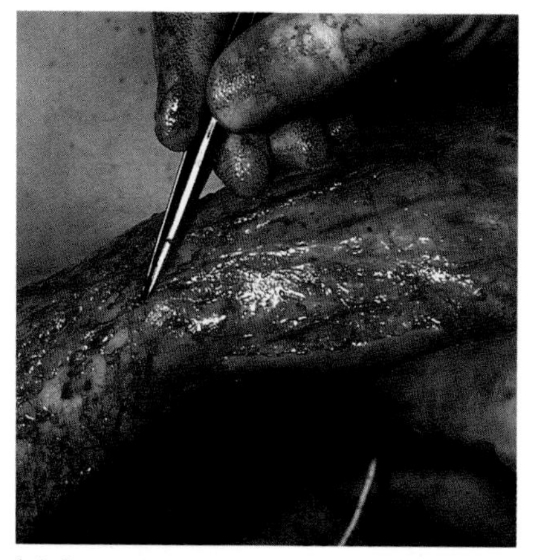

(c) Superficial veins are patent.

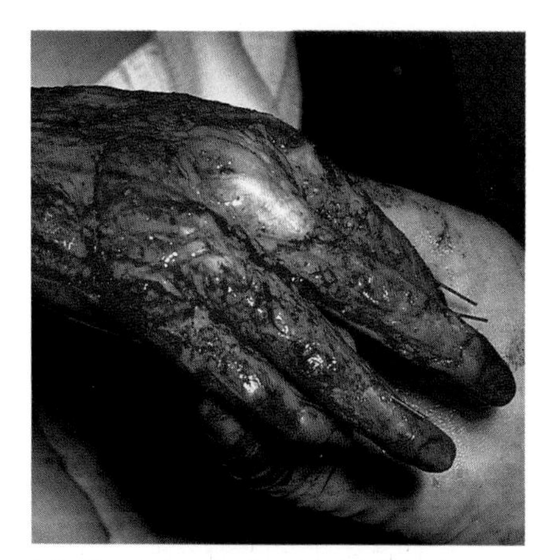

(d) The interphalangeal joints have been fixed in extension with fine Kirschner wires.

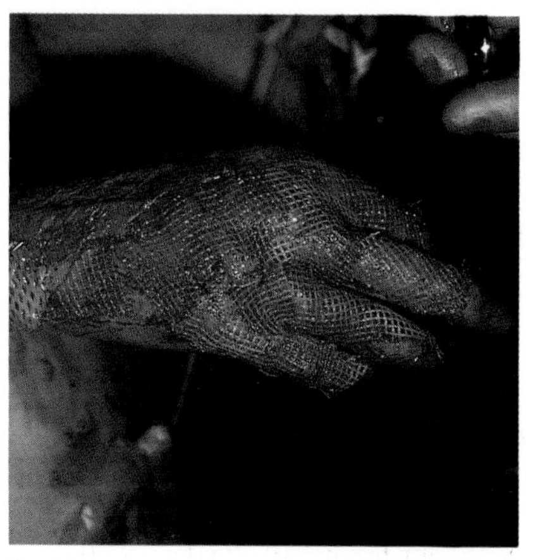

(e) The wound was subsequently covered with meshed split skin grafts.

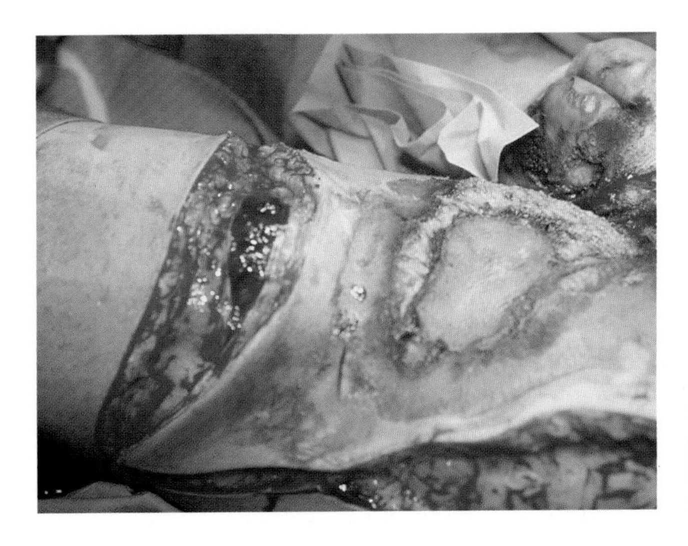

Fig. 5.13 Fasciotomy. A very deep electrical burn to the groin, with a large tissue deficit. A fasciotomy was performed to release and decompress the swollen muscles of the upper thigh.

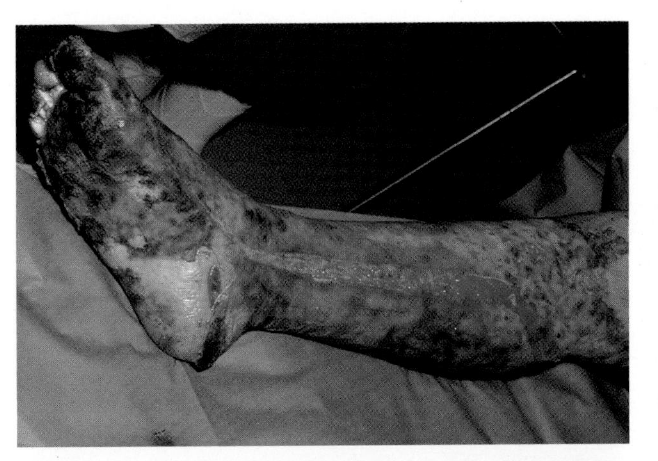

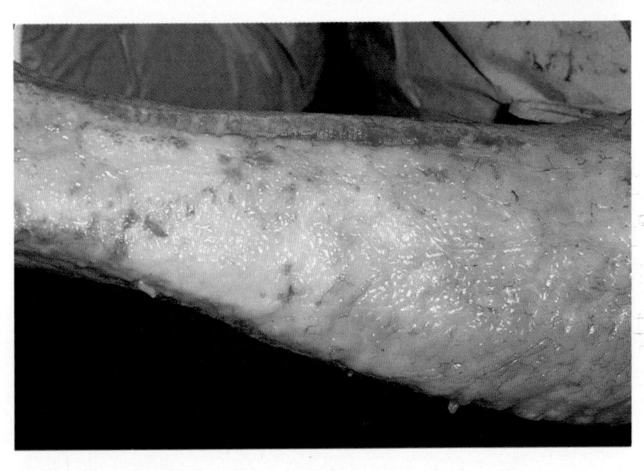

Fig. 5.14 Excision to fat. (a) A deep flame burn to the leg has been treated by exposure for three weeks.
(b) Excision under tourniquet revealed pockets of subeschar sepsis.
(c) Thorough excision to healthy fat has been performed.

Preparation

Surgery places considerable stress on the burned patient, therefore physiological and psychological preparation should be ensured. Surgery must be delayed wherever possible in patients with chest infections, serious sepsis or those who require artificial ventilation, where the outcome can be poor. The elderly and those with significant premorbid illness should be treated gently, and not submitted to aggressive surgery.

The wound should be carefully prepared, with the results of bacterial cultures readily available, enabling appropriate antibiotics to be selected for perioperative use. The general condition of the patient should be improved as much as possible; haemoglobin and other biochemical parameters should be checked and be near to normal. Preoperative sedation and analgesia should be administered as appropriate.

Surgery results in rapid heat loss in the patient. The operating theatre should be kept warm, at around 30°C. Further heat loss is prevented by covering and, wherever possible, placing the patient on a heated mattress. All fluids for transfusion, and anaesthetic gases are warmed before administration.

Ideally, all surgery and immediate post-surgical procedures should be rapid. Teamwork is therefore vital (Fig. 5.15). It may be necessary to move the patient repeatedly, in order to enable changing of drapes, cleaning of the wound area and application of dressings. As blood loss may be massive, large volumes of blood should be readily available before surgery commences.

Venous access is critical: at least one large bore cannula should be fixed into place (Fig. 5.16). Special care must be taken not to dislodge lines or the endotracheal tube when moving the patient.

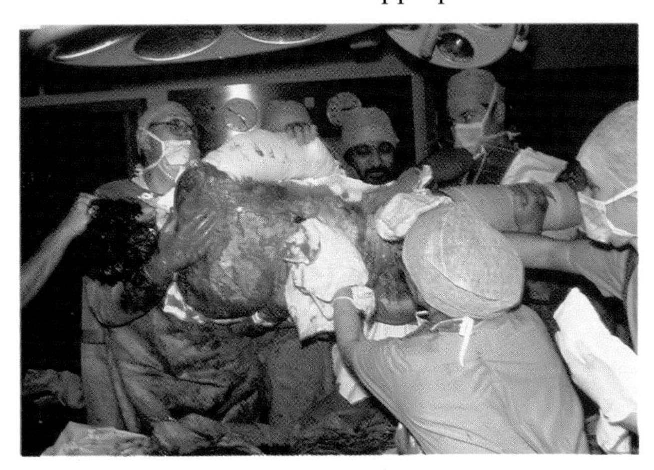

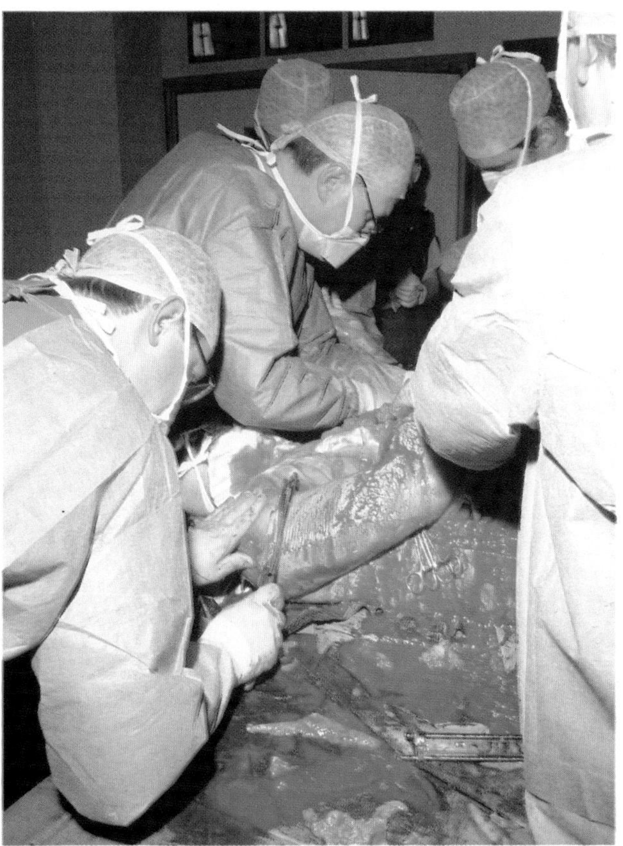

Fig. 5.15 Teamwork. (a) All attending staff must be prepared for surgical procedures: moving and cleaning the patient, changing drapes and applying dressings. All changes must be performed quickly and smoothly.
(b) Blood loss may be massive and almost uncontrollable.

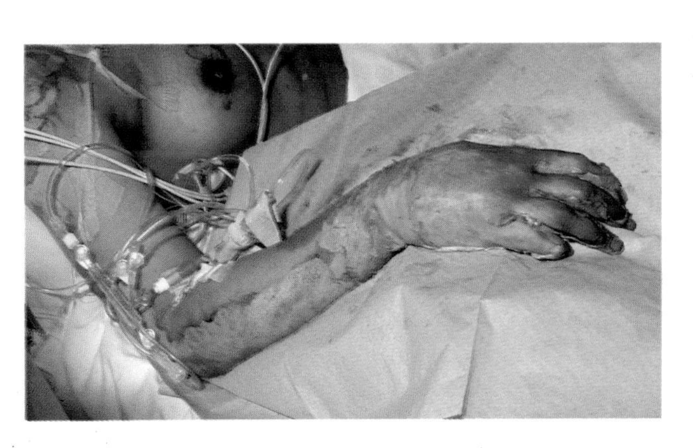

Fig. 5.16 Large intravenous cannula. Fluid replacement, parenteral nutrition, antibiotics and other drugs are administered through a central venous line.

Technique

Dressings are quickly removed from the patient and gloves and gowns of all attending staff are changed. The skin and burn wound is thoroughly cleaned with a topical antiseptic and subsequently dried. The patient is then placed onto a clean and impermeable sheet.

In extensive burns, a decision must be made on which surface is to be excised first and by which method. In circumferential burns of the trunk, either the anterior or posterior surface and upper thighs may be excised at one time. Excision is carried out with a skin grafting knife, opened wide. Limbs may be successfully excised under tourniquet. The wounds are covered with hot saline packs and elevated before the tourniquet is removed. Other areas should be excised promptly; all bleeding points are occluded with a fine catgut stitch. As cautery is usually unsuccessful, diffuse oozing is controlled by the application of hot wet packs and pressure.

Bleeding is always difficult to control, and is more profuse at superficial levels of excision and in the inflamed layer of a burn wound, or within fatty tissue. Dermal bleeding usually ceases quickly and spontaneously. Control of bleeding points may be gained by excision, using sharp dissection with a scalpel at the level of the deep fascia, where perforating vessels can be controlled by ligature or diathermy. Alternatively, a 0.5cm margin at the wound edge can be left unexcised as bleeding can be profuse in this area. However, all dead and damaged tissue should normally be completely removed, as any that is remaining and is incorporated in a dressing or is lying beneath a graft will act as a nidus of infection.

On completion of excision, blood clots and debris are rinsed off and the wounds covered with hot wet packs and bandages. Gloves, gowns and drapes are changed again in order to begin skin grafting.

WOUND CLOSURE

Priority areas

It may not be possible to cover extensive burns with skin grafts at the first operation, thus functional areas such as hands, neck, elbows, knees and feet should be given priority. Early healing of the hands gives the patient some independence, and rapid cover of the neck helps to prevent contracture and subsequent difficulties with anaesthetics. Elective skin coverage of joints prevents contracture and speeds mobilization. Grafting of the face and the back may be delayed as a significant amount of spontaneous healing will occur. It is wise and advisable to attempt to have the anterior surface of the body soundly healed before turning attention to the back.

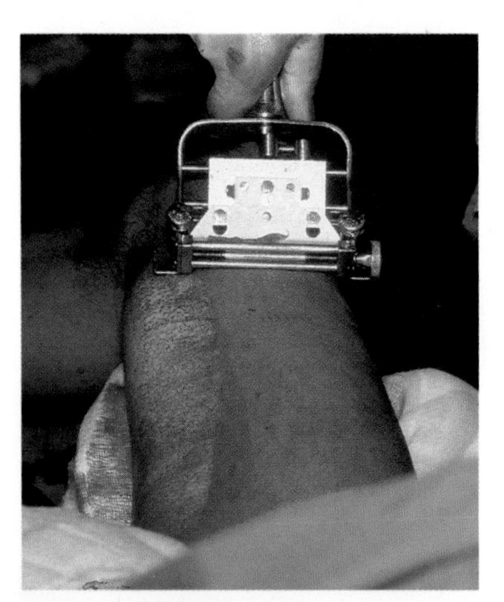

Fig. 5.17 Donor sites. (a) The thighs are prime donor sites as they offer large areas of skin, are convenient to use, and can be easily dressed.

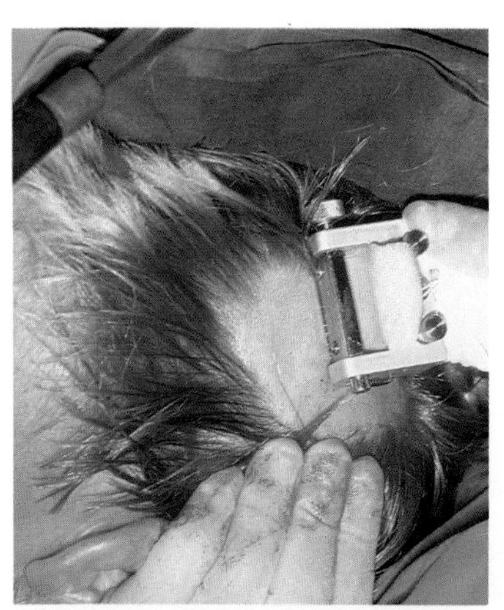

(b) The scalp is also an excellent donor site, particularly for the face, as the skin is a good colour match.

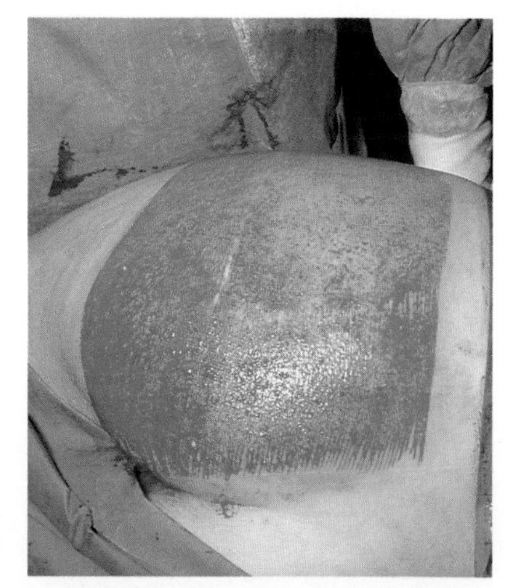

(c) The buttock as a donor site. Any scarring as a result of this should be relatively unnoticeable.

Donor sites

The availability of donor sites is dependent on the size and distribution of the burn. Skin is more easily taken from the thighs and calves, and is more difficult to crop from the abdomen and chest wall (Fig. 5.17). The scalp is an excellent donor site: it heals very rapidly and is relatively comfortable for the patient. Skin can be cropped three or four times before reaching the hair follicles; other sites may require two or three weeks for sufficient healing to occur before another graft may be taken. Where a small amount of skin is required, for example in children and young women, the buttock is an excellent site for donor skin. Any scarring will be relatively unnoticeable.

A description of the application of dressings for donor sites is seen in Fig. 5.18. It is important that the adherent dressings are not forcibly removed before two weeks.

Skin grafts

Grafts may be applied as sheets, strips, stamps or mesh. Meshed skin grafts are often used (Fig. 5.19). The skin can expand several times in size, and holes in the graft allow the free drainage of blood, serum or pus. This enhances graft take and accelerates wound closure.

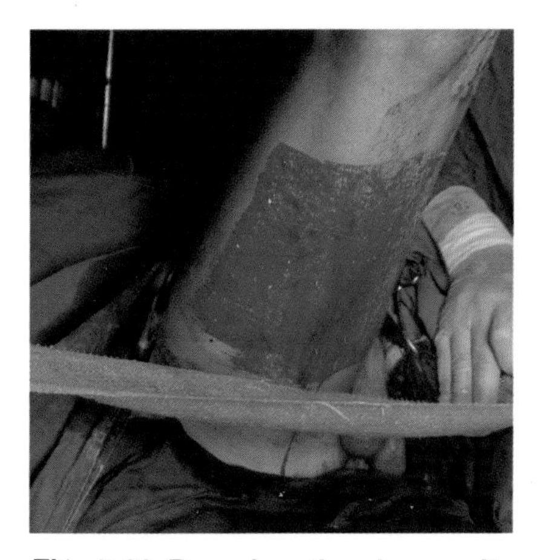

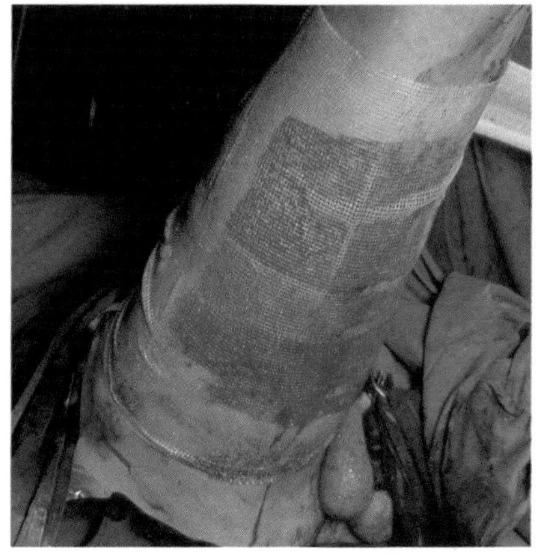

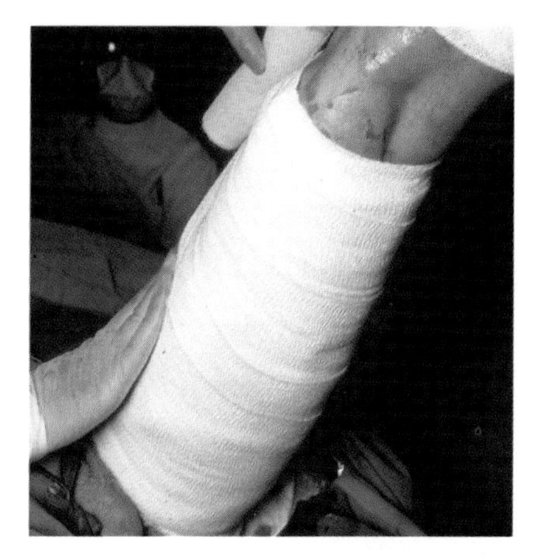

Fig. 5.18 Dressing the donor site. (a) Skin has been harvested from the thighs.

(b) Paraffin gauze is applied.

(c) Cotton gauze, cotton wool and a firm elastic crepe bandage are also applied and kept on for at least two weeks. They are then removed by soaking.

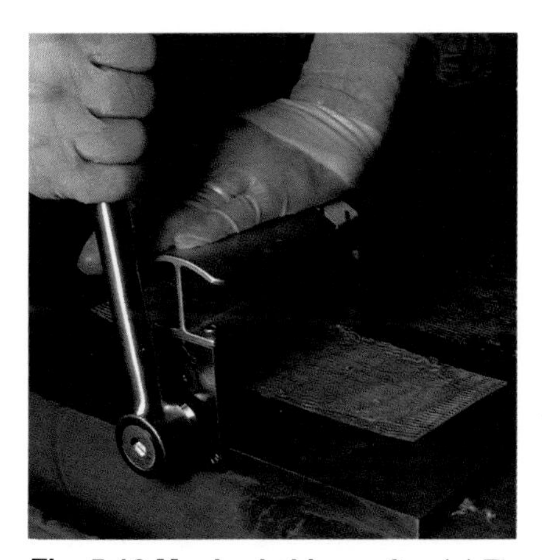

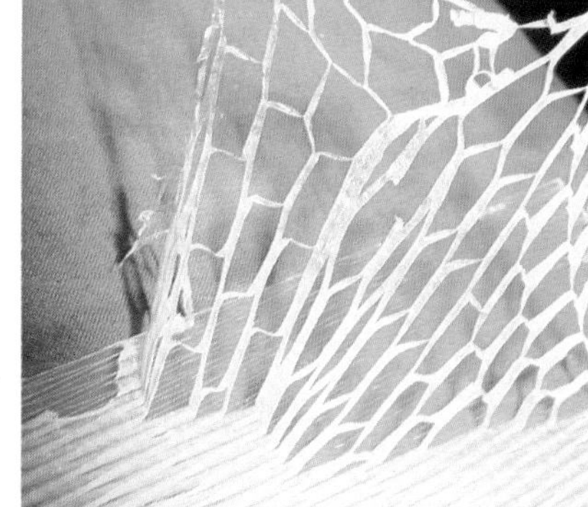

Fig. 5.19 Meshed skin grafts. (a) These are prepared by passing split grafts through a mesher which cuts perforations into the skin.

(b) This allows easy expansion.

(c) Meshed skin grafts are particularly useful in covering freshly excised or infected granulating wounds.

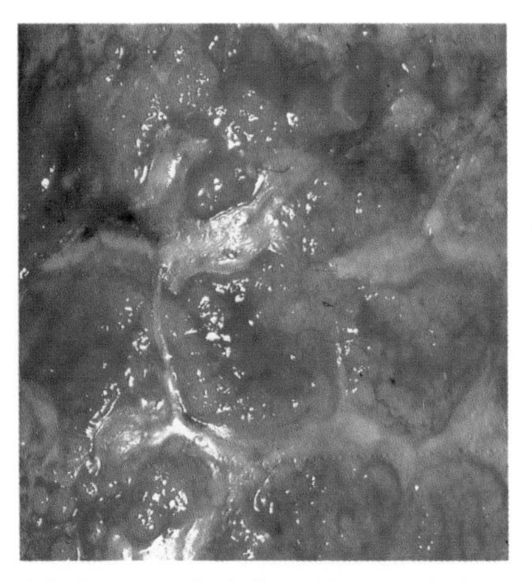

(d) On heavily infected granulating tissue, much of the graft will take; the interstices slowly close as the infection is brought under control and the condition of the wound bed improves.

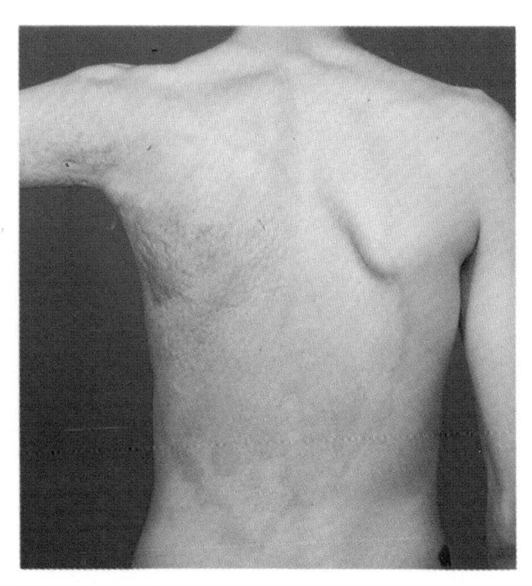

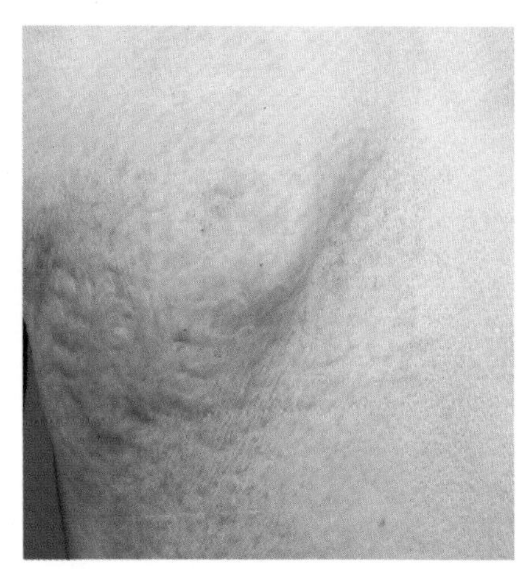

Fig. 5.20 Cosmetic appearance of meshed skin graft. (a) A 6:1 expanded graft, nine months after the injury. (b) Close-up appearance showing the pattern of the mesh.

Fig. 5.21 Taking of grafts. Split skin grafts should be taken thinly to facilitate rapid healing of donor sites.

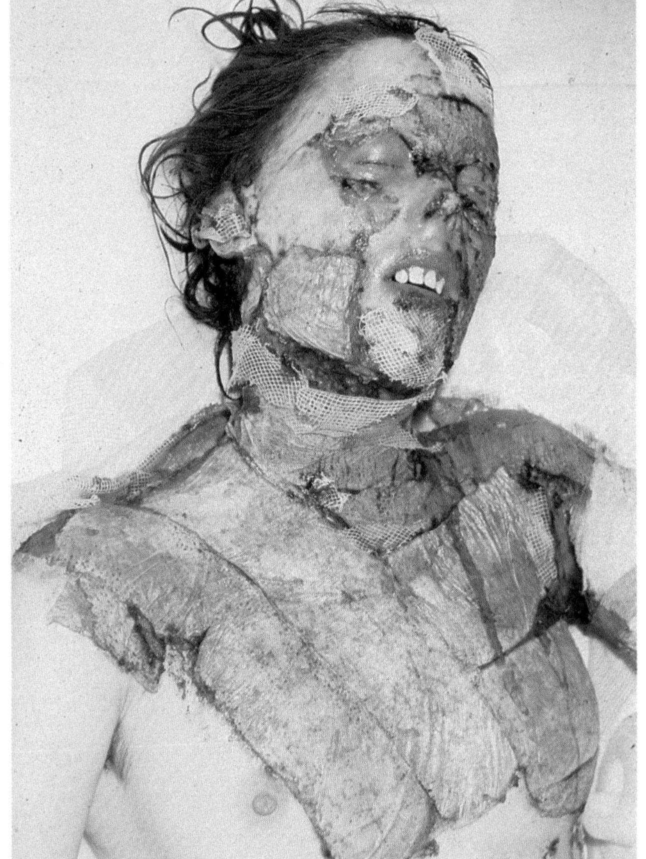

Fig. 5.22 Split skin grafts. (a) Thin sheets of split skin graft are spread, raw surface uppermost, onto paraffin gauze. The gauze is trimmed to fit the sheet precisely. (b) When split skin grafts take completely, they give the best cosmetic results and should be used in exposed areas such as the face.

However, as the pattern of the mesh persists (Fig. 5.20), this method ideally should not be used for exposed areas such as the face and hands. Several commercial skin graft meshers are available permitting expansion from 1.5:1 to 9:1.

Split skin grafts should be taken in a thin sheet to enable the donor site to heal quickly (Fig. 5.21). A hand-held dermatome is the most convenient instrument for normal graft-taking, although a mechanical dermatome is more suitable in harvesting larger areas of skin.

Sheet grafts are preferably placed on excised, non-granulating wounds. These sheets of skin are initially spread and trimmed on a layer of paraffin gauze and applied directly (Fig. 5.22). If suitably moistened, they may be stored in a refrigerator at 4°C for up to three weeks. The wound bed must be clean, with no bleeding. The cosmetic results of any grafting are usually poor and large sheets of thick skin should ideally be reserved for reconstruction.

Strips or postage stamp-sized sheets of skin are usually used to cover heavily contaminated surfaces, as drainage is not impeded; secondary epithelialization from the edges of the graft is also rapid.

Graft dressing

The decision to apply a dressing is dependent on the type of graft. Sheet skin grafts may be left exposed, without a dressing whereas meshed skin grafts should not be allowed to dry as subsequent crust formation in the interstices delays wound healing and closure.

The normal graft dressing consists of vaseline-impregnated gauze, several layers of cotton gauze, a layer of cotton wool and a firm bandage to immobilize the area. Dressings are kept on for five days; after this time the outer layers are gently removed down to the vaseline gauze and clean dressings reapplied every two or three days until healing is progressing soundly. If the wound is dirty and infected, the vaseline gauze should also be replaced.

Meshed skin grafts on infected wounds should be dressed more frequently, and if necessary, covered with soaks of hypochlorite or silver nitrate solutions. Care must be taken not to dislodge the graft, although some displacement may be inevitable. Surfasoft , a permeable plastic interface which is placed over the meshed skin graft, provides considerable protection. It is non-adherent to the underlying skin provided the gauze dressings are kept moist. Dressings can be changed as frequently as indicated.

Special techniques

In large burns (particularly in children) where extensive areas must be covered, with only a limited amount of skin available for grafting, mixed meshed autograft and allograft applied as a 'sandwich' can be a valuable technique (Fig. 5.23). Alexander and McMillan introduced the technique of using a widely-expanded mesh graft of the patient's skin, covered with a less expanded mesh graft of donor skin. Both grafts 'take', with the allograft serving to protect the autograft against dessication and bacterial colonization, while the autograft commences epithelialization. However, allograft skin can transfer the viruses responsible for hepatitis B and AIDS, and this method should only be used in extreme cases.

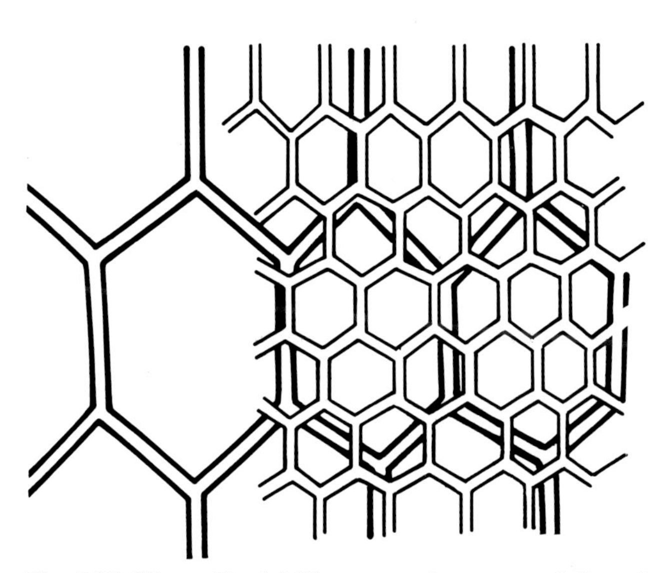

Fig. 5.23 Allografts. (a) Diagrammatic representation of the Macmillan model.

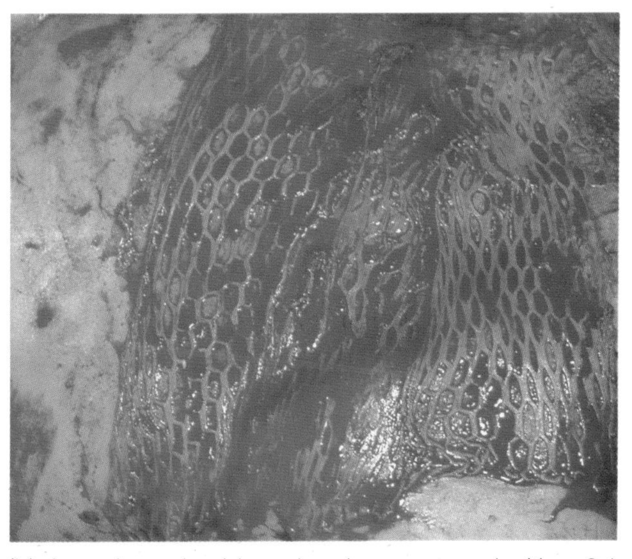

(b) A newly excised burn has been covered with a 6:1 meshed autograft.

The same technique can be used with glycerol-preserved allograft skin, fresh xenograft (normally pig skin) or fetal membranes (Fig. 5.24), used as the allograft. Artificial skin substitutes are also available which provide temporary wound closure before definitive closure with a skin graft is carried out, but these are of little use in the presence of infection.

Cultured keratinocytes for use in grafting have enjoyed a passing popularity (Fig. 5.25) but have been found to be too delicate to provide rapid wound closure or prevent severe scar contracture. Research is now being directed towards the development of artificial membranes which would provide a scaffolding for the semi-reconstruction of dermis, which in turn could support a thin epithelial graft or cultured keratinocytes. Such materials are currently under trial.

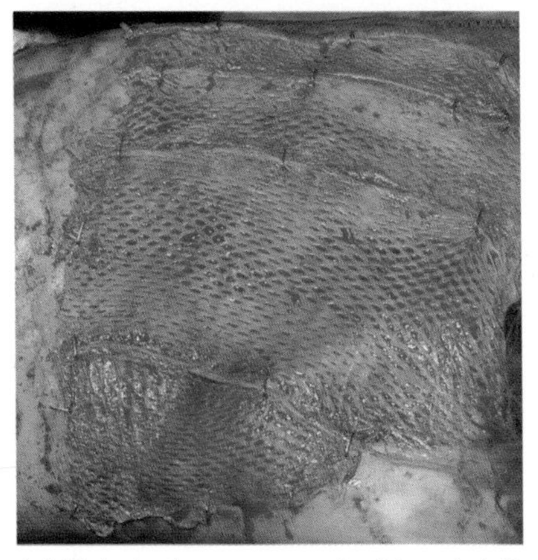

(c) This is then covered with a 1:1.5 allograft, taken from the child's father.

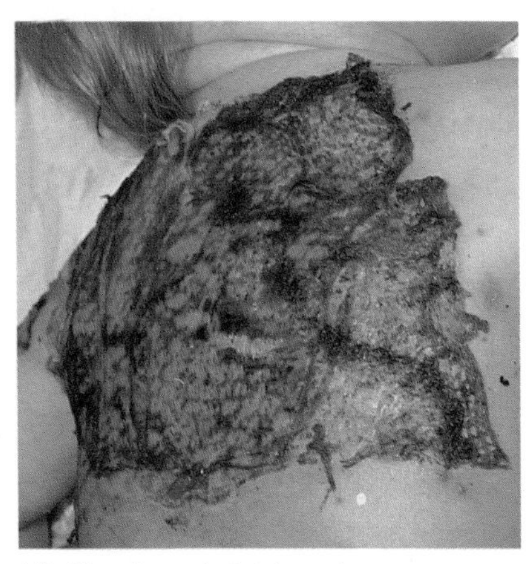

(d) The 'sandwich' took completely, with no graft loss from infection or rejection.

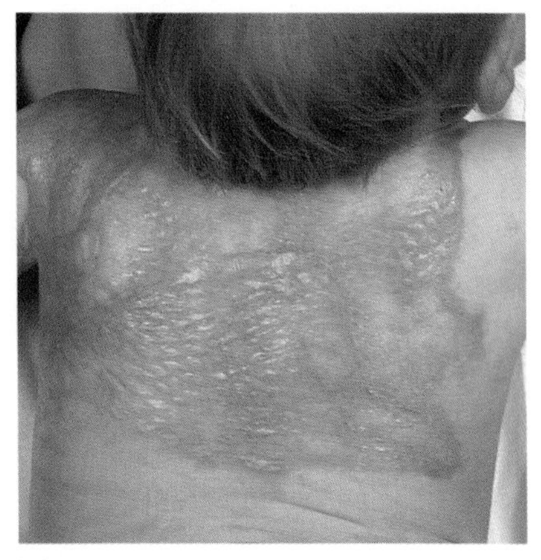

(e) Healing proceeded with no further treatment required.

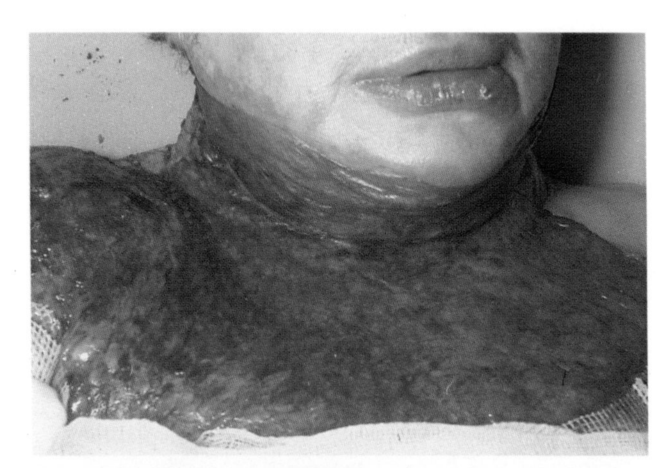

Fig. 5.24 Fetal membranes. (a) These have been applied to a deep dermal burn.

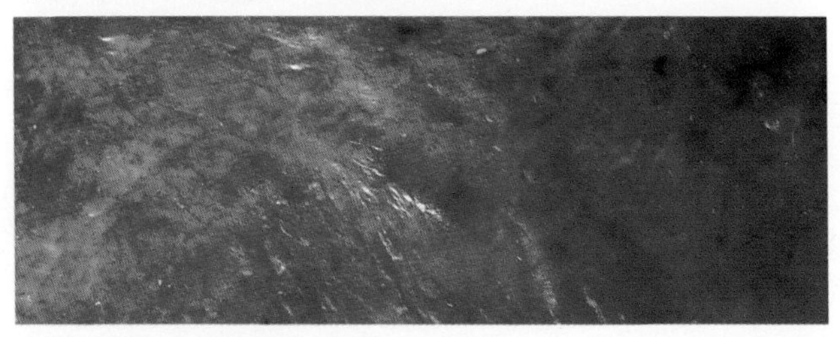

(b) Appearance close up.

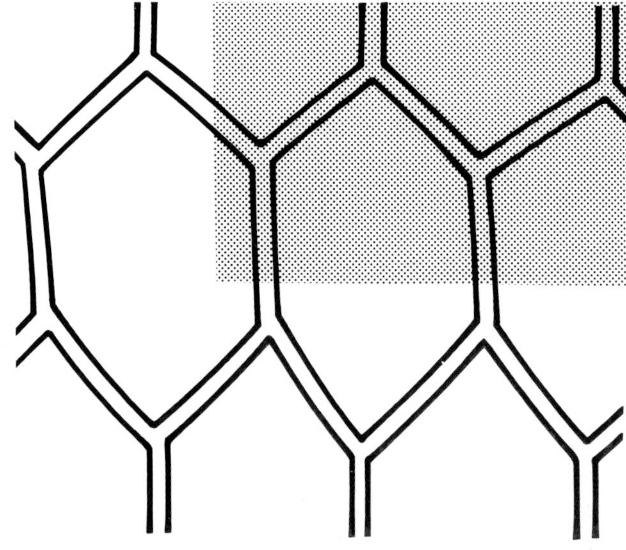

Fig. 5.25 Application of cultured epithelium. (a) Diagrammatic representation of meshed skin graft with cultured epithelium.

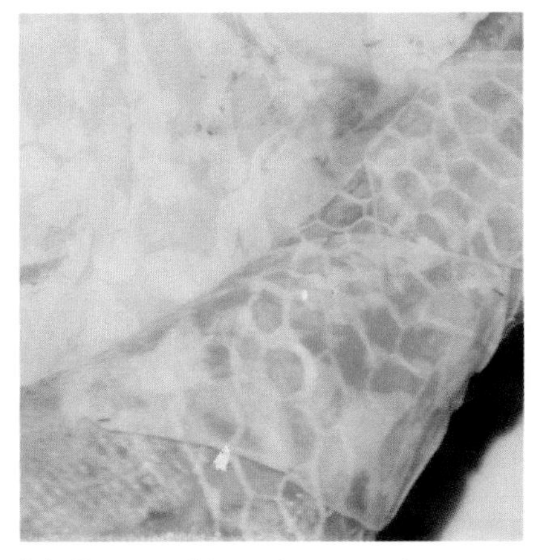

(b) Sheets of keratinocytes have been applied to widely expanded meshed skin.

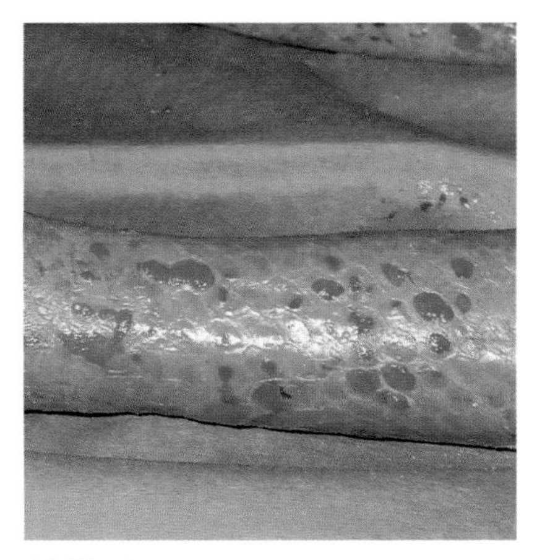

(c) The interstices have closed quickly.

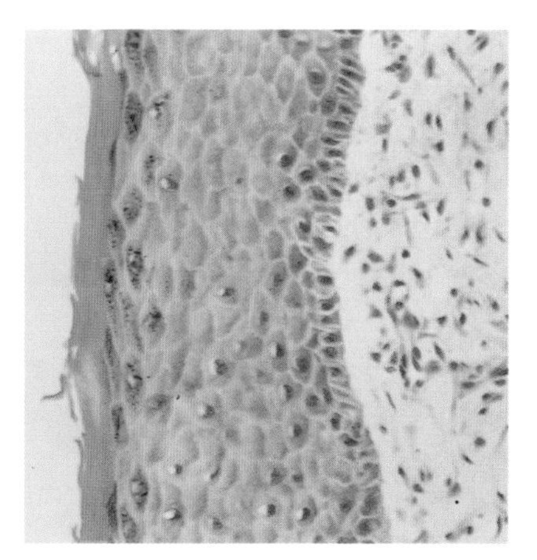

(d) A biopsy demonstrates formation of a new skin layer.

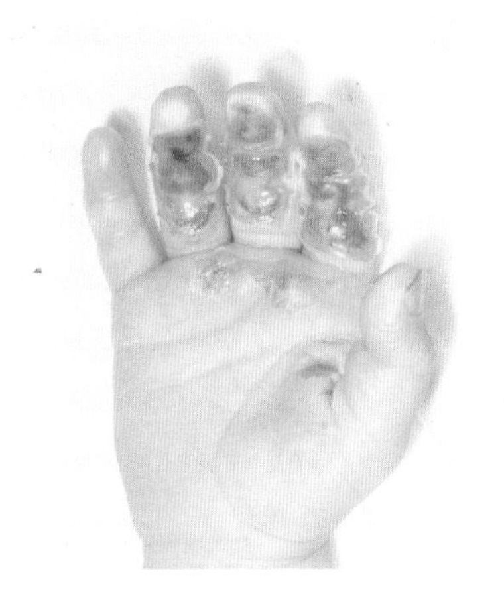

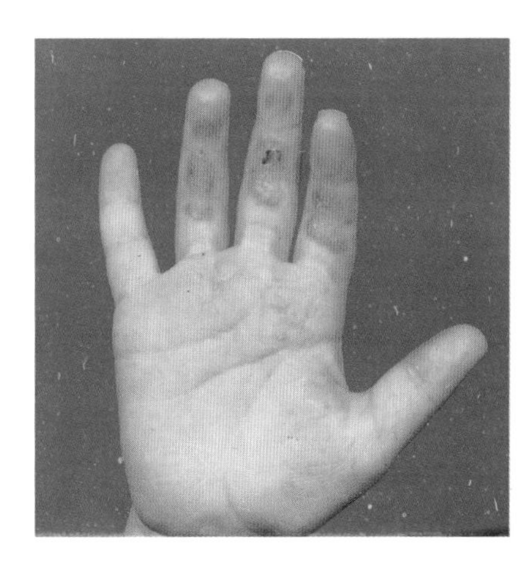

Fig. 5.26 Treatment with silver suphadiazine cream in a plastic bag. (a) Small, deep contact burns on presentation.
(b) The burns have healed spontaneously after four weeks and any contracture can be dealt with after a period of splinting.

BURN INJURIES OF SPECIAL AREAS

Hands

Treatment of superficial and dermal burns of the hands using a plastic bag containing antiseptic cream will permit rapid and painless mobilization, and also provide the best environment for epidermal regrowth and dermal preservation (Fig. 5.26). If the burn has not healed within 20 days, grafting should be seriously considered.

In deep burns, where spontaneous re-epithelialization is unlikely, vascular granulation tissue eventually develops which cannot protect tendons, ligaments or joints from permanent damage. This may affect the hand in several ways. Tendons will weaken, cease sliding and thus become adherent, and the ligaments surrounding the joints shorten, holding the joints in a contracted position. The joint capsule itself contracts with shortening of the adjacent intrinsic muscles, resulting in the gradual hyperextension of the metacarpophalangeal joints and flexion of the interphalangeal joints in the fingers (Fig. 5.27). Attempts to improve the position of the wrist and fingers by splinting inevitably places pressure over the extensor surfaces, further damaging the extensor tendons and joint capsules of the proximal interphalangeal joints. Following excision and grafting, the hands should be splinted using a volar slab to maintain the wrist in 60° extension, the metacarpophalangeal joints in 90° flexion and the interphalangeal joint in full extension. In this position the joint ligaments are stretched and restoration should be straightforward. Fig. 5.32 shows how difficult this is to achieve in practise: if the dressing is bulky, the position of the hand is more likely to become displaced.

In order to avoid these developments, the early excision of deep burns is performed and a skin flap or graft is applied promptly (Fig. 5.28). Skin flaps can be taken from sites such as the upper arm or groin (Fig. 5.29). The fingers can be temporarily immobilized by passing a Kirschner wire across the interphalangeal joints until the grafts have healed. This course of

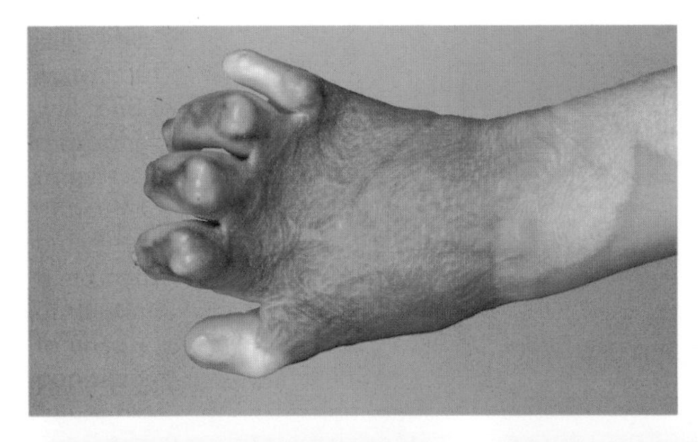

Fig. 5.27 Hyperextension of the metacarpophalangeal joints. The intrinsic muscles in the hand have shortened and the metaphalangeal joints have become hyperextended, with the fingers flexed.

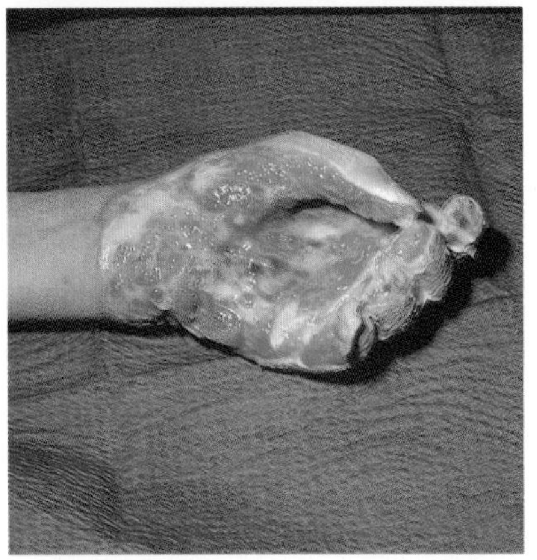

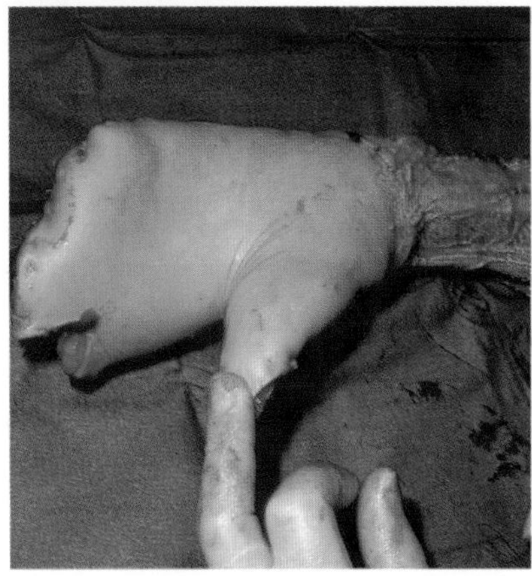

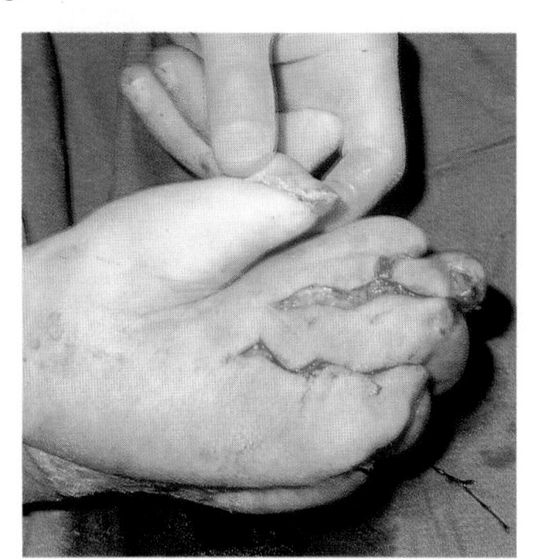

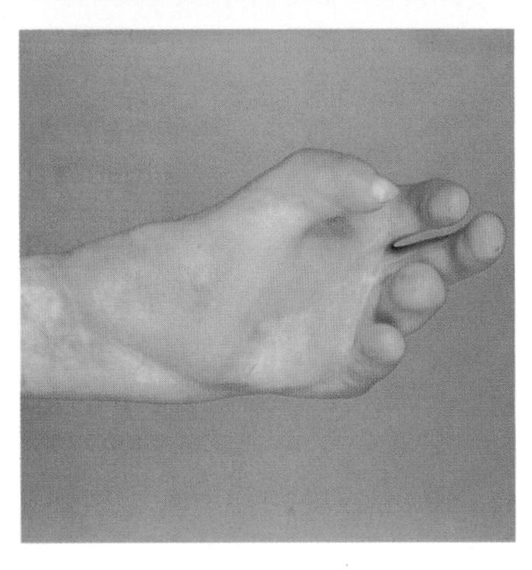

Fig. 5.28 An extremely deep, self-inflicted burn of the hand. (a) This injury involved the palmar surface only. (b) The burn has been excised and covered with a reversed radial forearm flap. The radial artery has been divided proximally and most of the skin of the anterior surface of the forearm raised and swung around on the arterial pedicle in order to cover the raw area. (c) The fingers have been syndactylized, and the donor site will be grafted. (d) Appearance four years later.

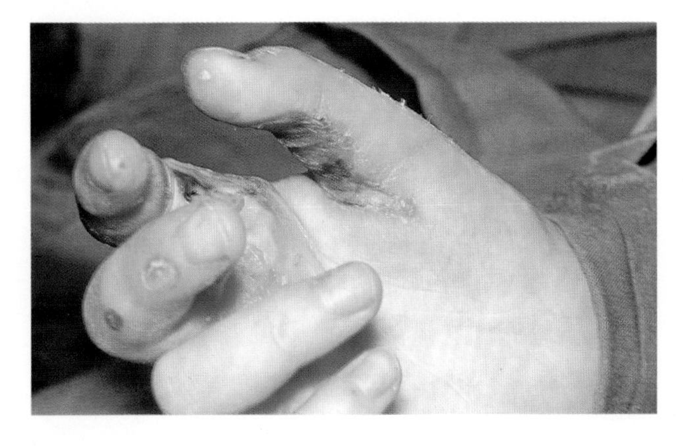

Fig. 5.29 Flap cover. (a) Small deep, electrical burns involving several digits.

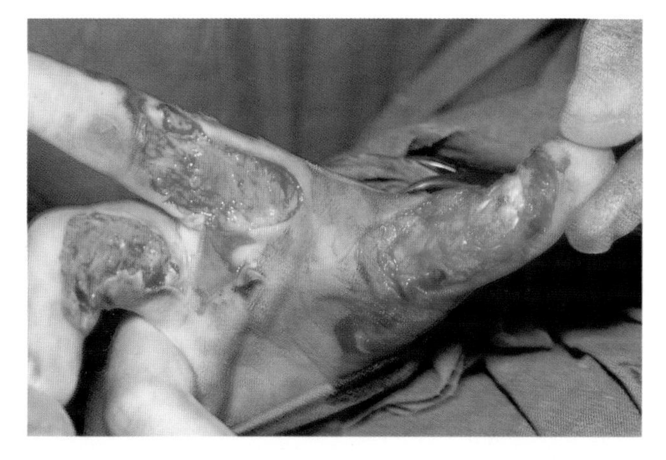

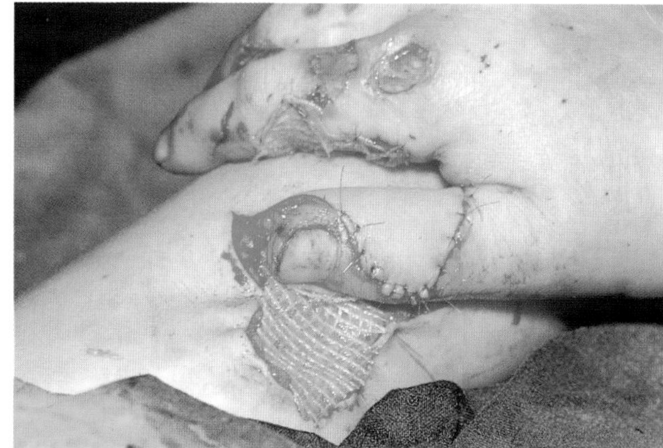

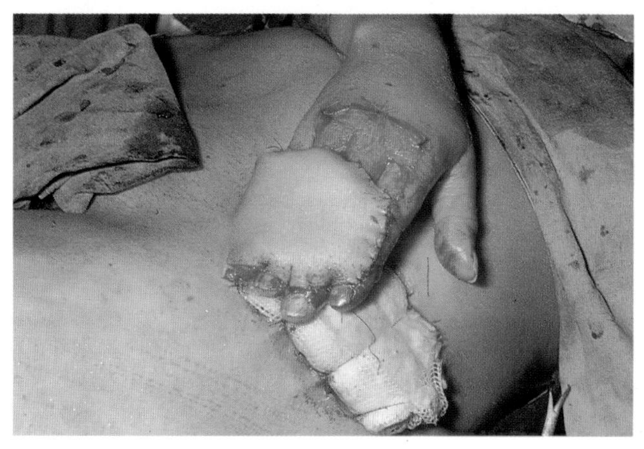

(b) The wounds have been excised.
(c) The area is covered with small flaps raised on the upper arm. The secondary defects are skin grafted.
(d) Groin flap. Deep burns involving the dorsum of the fingers, exposing the extensor tendons, have been covered with a groin flap. The flap was peeled away at three weeks, leaving behind a bed of healthy fat which was skin grafted.

treatment is applicable in isolated burns of the hand; however, in extensive burns of the body other aspects may take priority and a compromise in treatment will be reached. Initial blistering with loss of the fingernails usually indicates thrombosis of the neurovascular bundles of the fingers. Amputation through the proximal interphalangeal joint will eventually be necessary (Fig. 5.30). In very severe burns of the hand, all necrotic tissue is excised immediately and covered with a skin flap (Fig. 5.31). Until surgery can be performed, burns of the hand should be treated in a plastic bag.

Face

Facial tissues are very well vascularized and have a good supply of hair follicles, sweat and sebaceous glands. The potential for spontaneous healing is therefore high. In cases such as flash burns, the wounds may heal as quickly as one week (Fig. 5.33). Most deep burns of the face can be treated by exposure and subsequently grafted (Fig. 5.34). However, in severe facial burns reconstruction of features is likely to be necessary using various techniques such as a tubed pedicle (Fig. 5.35) and skin grafts bearing hair (Fig. 5.36). If wounds are treated by exposure, it is important that the surfaces are kept clean and free from a build-up of protein exudate.

Secretions from the mouth, nose and eyes keep the surrounding tissues moist, providing a hospitable environment for bacterial colonization (Fig. 5.37). Continuously growing facial hair will firmly trap any crusts which have formed and secondary inflammatory changes will take place in the wound bed. Increased build-up of granulating tissue will also worsen the hypertrophic scar contracture. Treatment is therefore directed towards the frequent changing of wet dressings of saline or chlorhexidine, together with regular removal of crusts and the trimming of facial hair. Alternatively, burns in this area may be covered with antiseptic cream which is removed by washing and reapplied twice daily. Again, the build-up of crusts must be actively treated. After any infection has been brought under control, raw areas can be grafted.

Fig. 5.30 Damage to the neuro vascular bundles, leading to amputation. (a) Considerable damage was seen to the thenar and web space muscles, and to the neurovascular bundles.

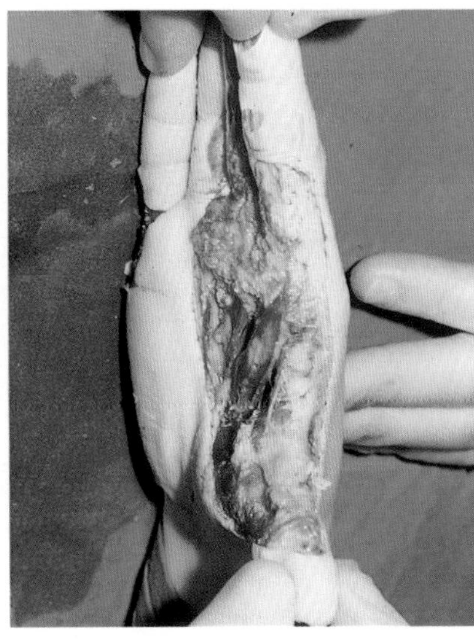

(b) The wounds have been excised.

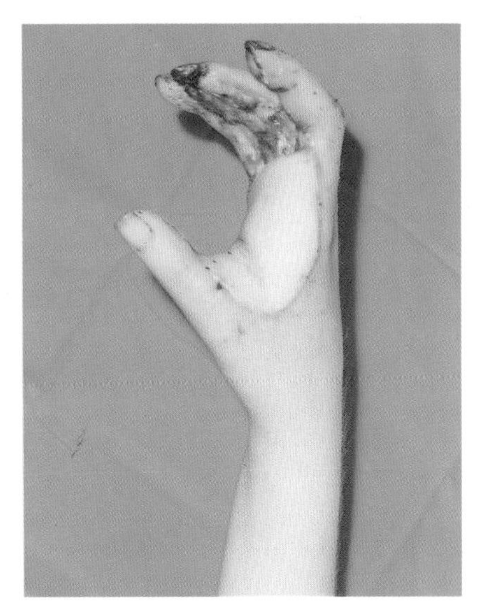

(c) The little finger was amputated and the hand covered with a groin flap.

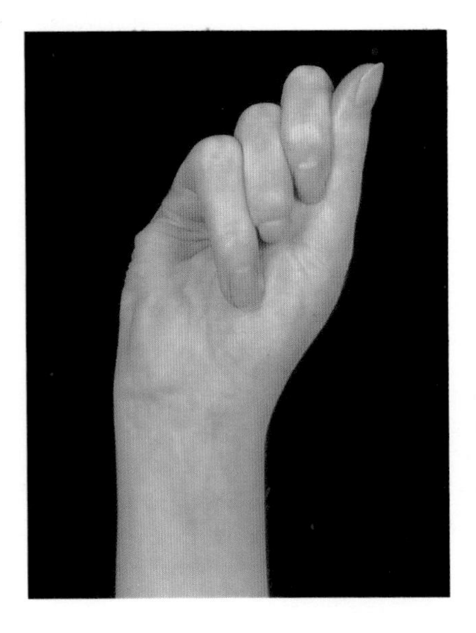

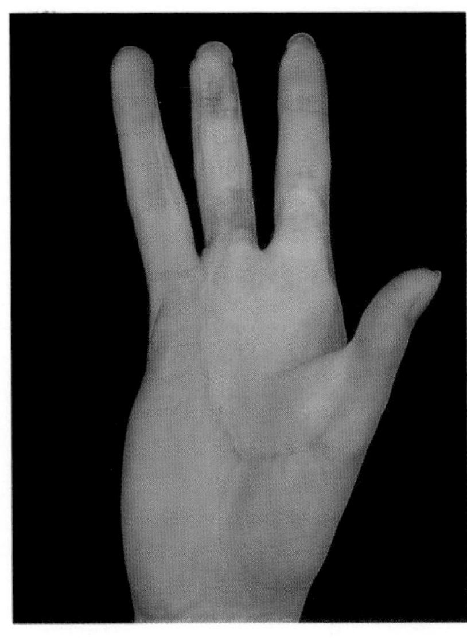

(d) Demonstration of function, one year later.
(e) Cosmetic appearance one year later.

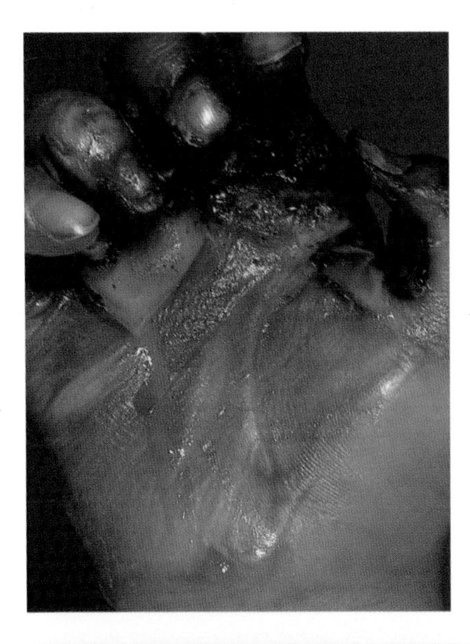

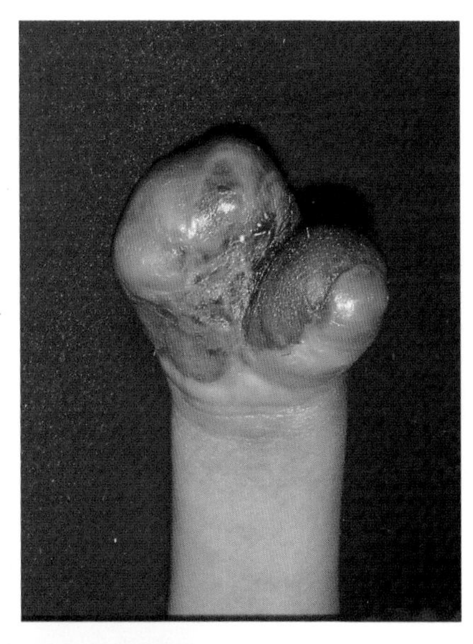

Fig. 5.31 Deep contact burns of the hand. (a) These injuries were sustained through contact with an electric bar fire.
(b) The necrotic tissue has been excised and the remnant buried in a groin flap for three weeks. The thumb-index web has been deepened and grafted to afford pinch.

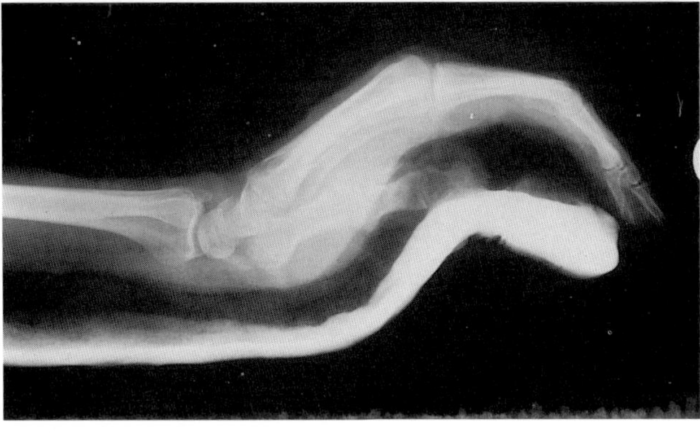

Fig. 5.32 X-ray of hand immobilized in a plaster slab.

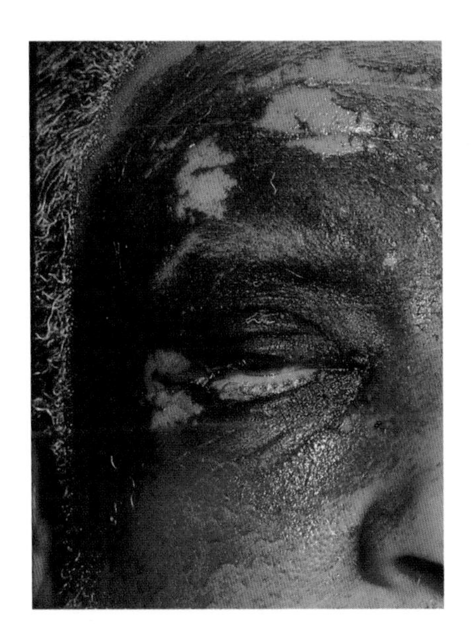

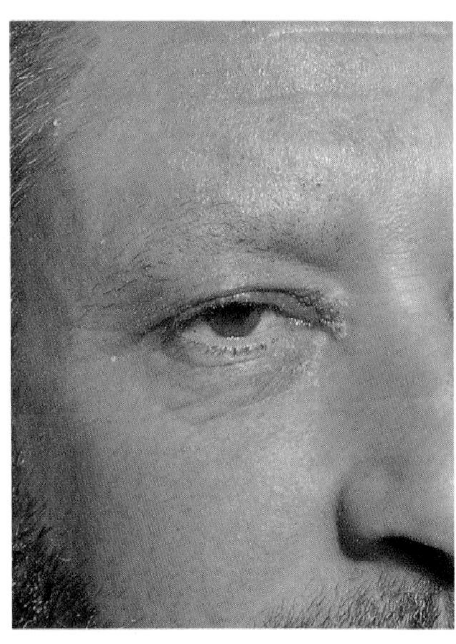

Fig. 5.33 Spontaneous healing of facial burns. (a) Injuries caused by high intensity electric arc of very high temperature but very short duration. The superficial layers are charred
(b) healing is complete at one week.

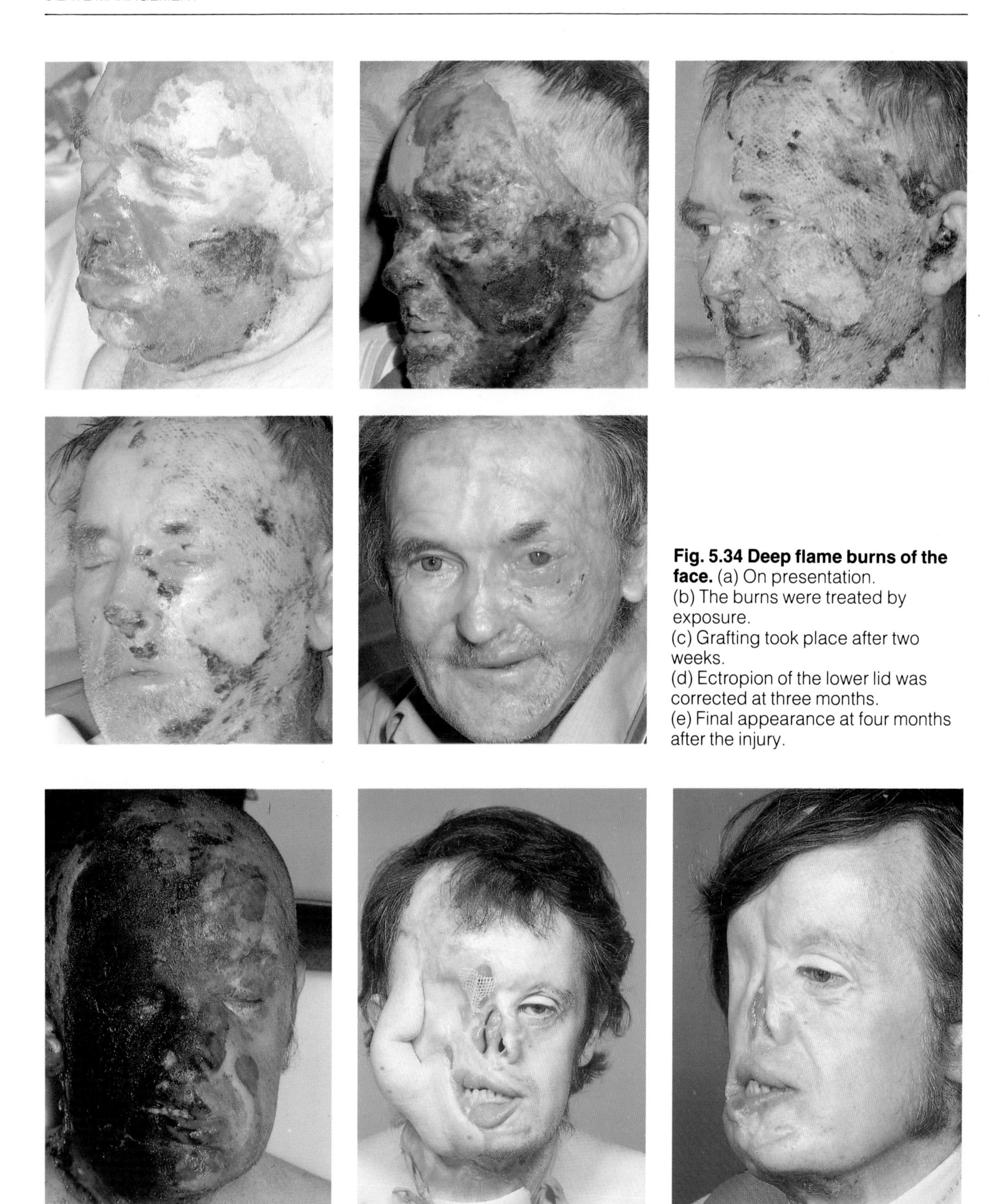

Fig. 5.34 Deep flame burns of the face. (a) On presentation.
(b) The burns were treated by exposure.
(c) Grafting took place after two weeks.
(d) Ectropion of the lower lid was corrected at three months.
(e) Final appearance at four months after the injury.

Fig. 5.35 Deep, charring burns of the face. These burns were sustained when the patient, an epileptic, fell into a fire. A long tubed pedicle was raised on the back and waltzed in slow stages to cover exposed jaw and skull. Reconstruction took nearly a year.

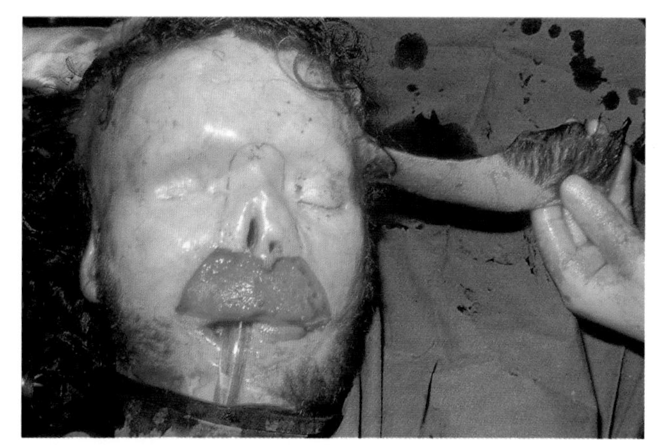

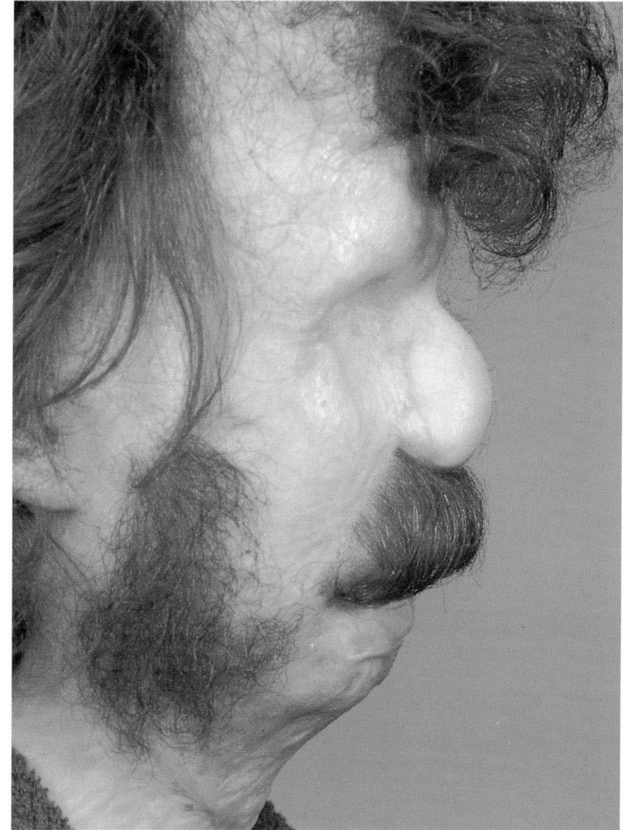

Fig. 5.36 Severe alkali burns of the face and eyes. (a) This has resulted in loss of most of the features. Reconstruction was carried out with a skin flap from the scalp, bearing hair, for the upper lip. Skin from the neck, also bearing hair, has been advanced to cover the jaw line.

(b) The nose has been constructed from a tubed pedicle raised from the inner arm.

Where burns of the face include the ears and neck, the patient should be nursed without pillows. This limits damage from pressure to the ears and facilitates hyperextension of the neck, thus helping to prevent contractures and enabling the area to be kept clean.

In burns of the mouth, the resultant thick hard inelastic hypertrophic scar tissue make feeding difficult and the introduction of anaesthesia dangerous. A simple device using an elastic band to maintain the tension can be used to spread the mouth at the commissures (Fig. 5.38).

Ears

In deep burns of the ear the risk of suppurative perichondritis, with progressive loss of undamaged cartilage, is high. The ears should be inspected frequently. Complaints of pain, with a red swelling anteriorly (Fig. 5.39) and loss of the normal angle behind the ear and mastoid indicates that perichondritis has developed. Pus has developed between the perichondrium and skin of the ear; as it expands, it strips and devascularizes the cartilage (Fig. 5.40). Every effort should be made to avoid local pressure and trauma, and dressings should be applied to keep the ear moist. Antibiotics are ineffective and the condition will progress rapidly and irreversibly unless

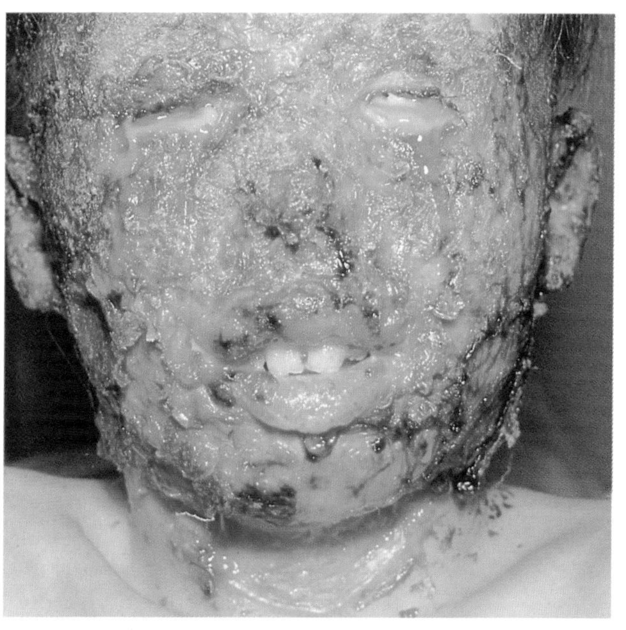

Fig. 5.37 Burns of the face which have been treated by exposure. Beneath the crusts, infected granulations will be found which are responsible for producing this thick crust.

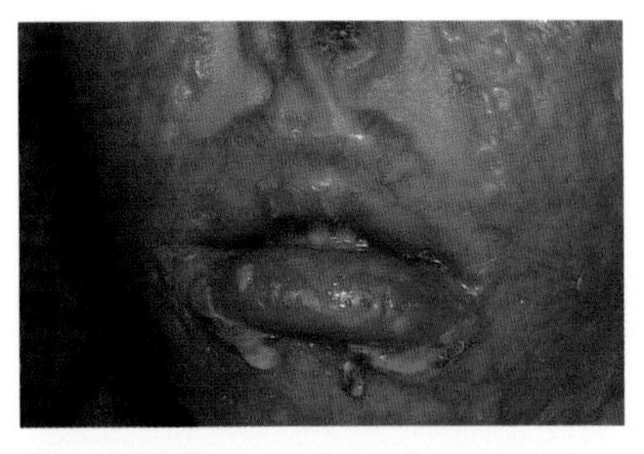

Fig. 5.38 Burns of the mouth. (a) Inelastic scars impede opening of the mouth.

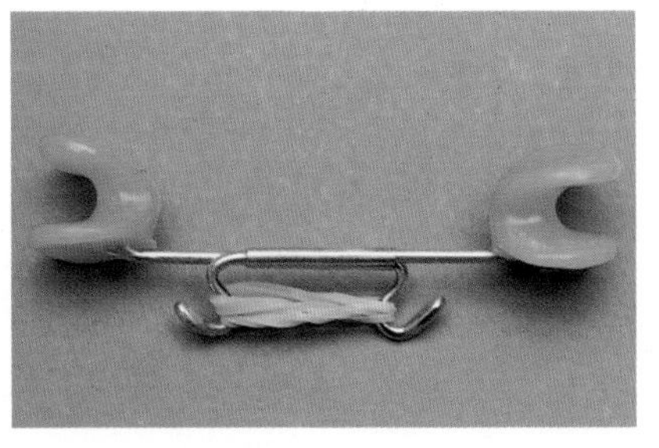

(b) A device can be used to spread the mouth at the commissures.

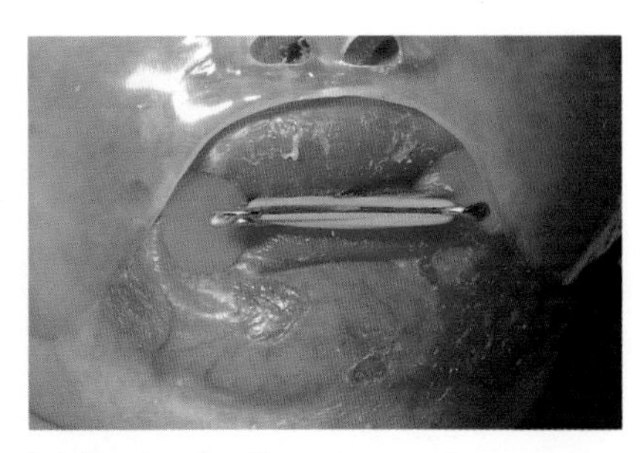

(c) Device in situ.

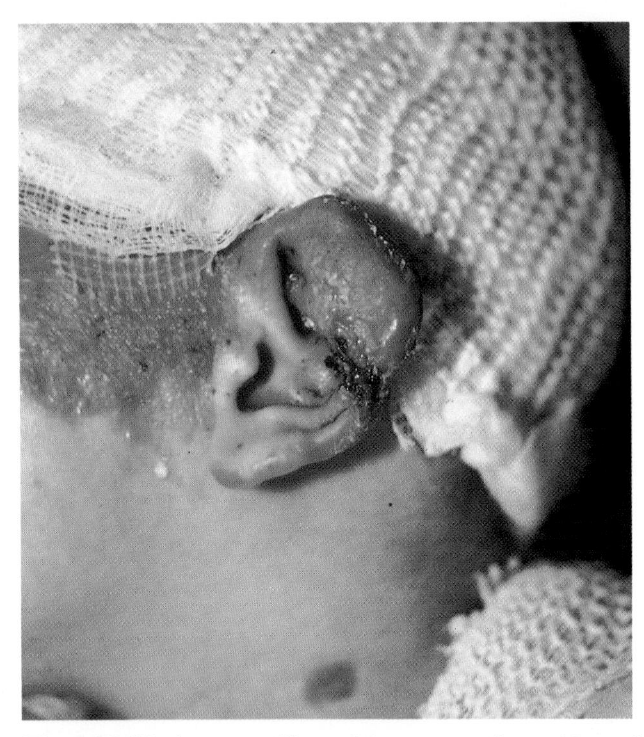

Fig. 5.39 Obvious swelling of the ear cartilage. Most of the cartilage beneath the swelling will have to be removed.

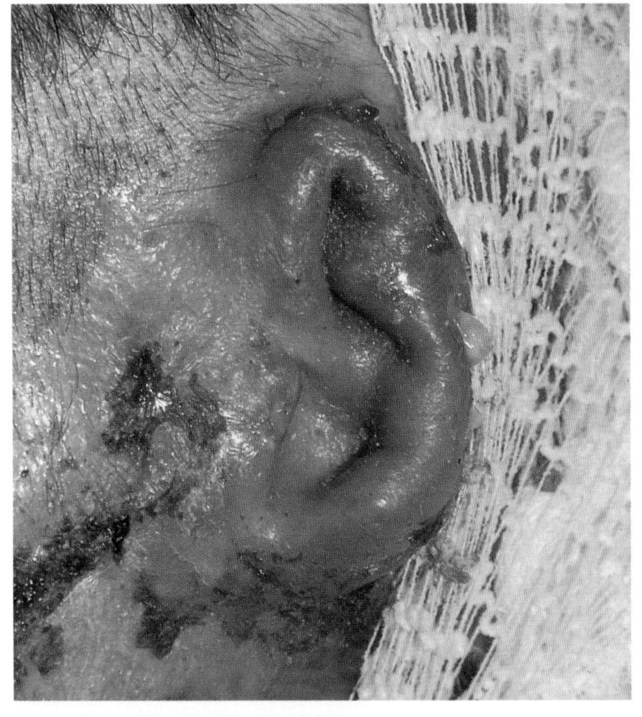

Fig. 5.40 Suppurative perichondritis. Exposure of the cartilage of the ear has resulted in an infection by *Pseudomonas aeruginosa*.

surgical decompression and excision of necrotic cartilage is performed. If these measures are not taken, the ear will become contracted and shrunken, and adequate reconstruction will not be possible (Fig. 5.41). Exposed, clean cartilage can be excised and grafted with minimal deformity.

Eyes

All chemical burns to the eyes should be treated in the same manner, i.e. with thorough and prolonged irrigation with water or normal saline (Fig. 5.42). Irrigation under general anaesthetic may be performed to suppress pain and panic. In acid burns to the eye, fluorescein staining will demonstrate any corneal damage whereupon an ophthalmic opinion must be sought. Injuries involving hot tar or bitumen results in adherence of the lids, usually without damaging the globe (Fig. 5.43). The patient should be reassured that sight will be undamaged. In burns resulting from an electrical arc flash, the blink reflex may be too slow to prevent damage to the cornea, although this is very unusual (Fig. 5.44).

The skin of eyelids is thin and easily damaged and the pull from surrounding facial scars can lead to secondary deformity. It is essential that the cornea is protected, even during anaesthesia. In this case, the cornea are covered with chloramphenicol eye ointment and protected with vaseline gauze (Fig. 5.45).

Any deformity of the lids, either primary or secondary, which leads to corneal exposure must be surgically corrected. The upper lid provides most

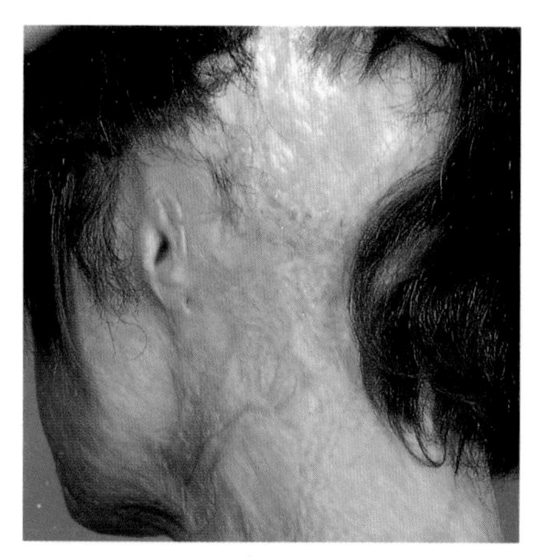

Fig. 5.41 End-stage perichondritis. The cartilages have been lost completely.

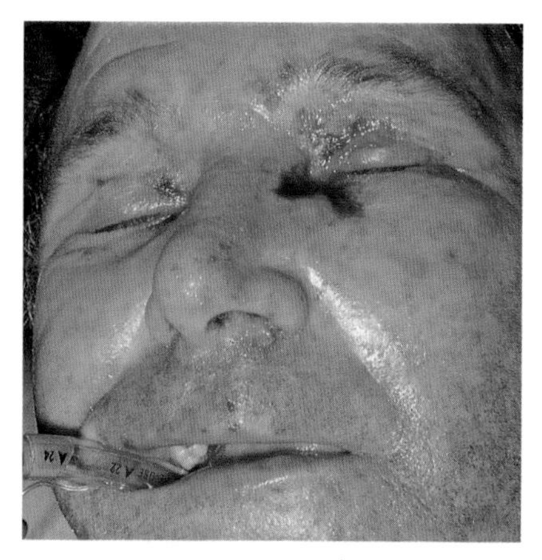

Fig. 5.42 Chemical burns to the eyes. (a) Strong acid was splashed onto the eyelids. Irrigation has been carried out.

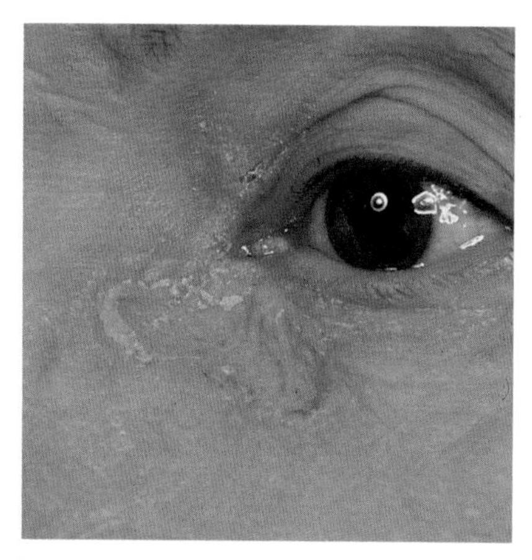

(b) Two weeks later.

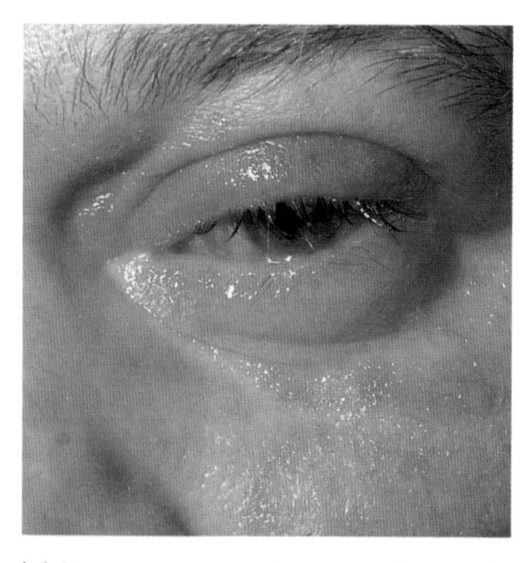

(c) In a separate patient, oedema of the lids and injection of the con junctiva can be seen following acid burns to the eyes; irrigation has been performed.

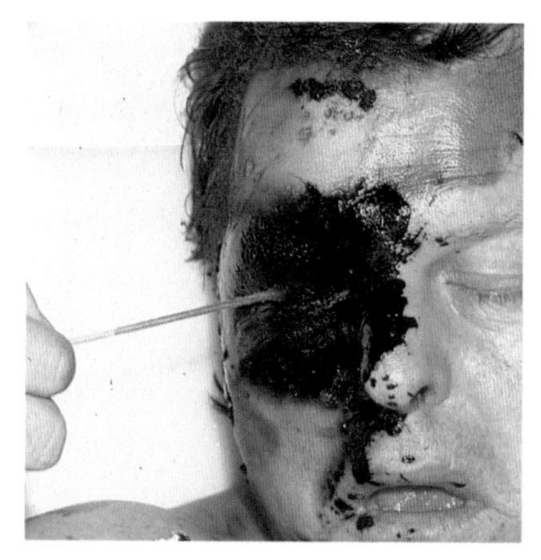

Fig. 5.43 Hot tar burns to the face and eyes. The lids are covered with vaseline ointment.

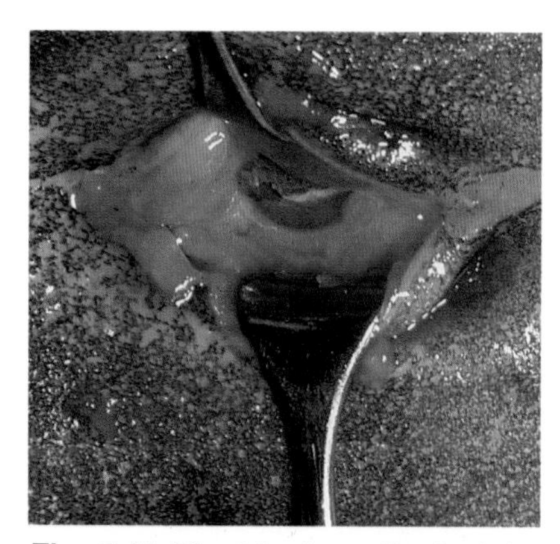

Fig. 5.44 Electrical arc flash. (a) Desquamation of the superficial layer of the cornea has resulted. This was wiped off and healed rapidly.

(b) Subsequent eversion of the upper lid with gross oedema of the conjunctiva (chemosis).

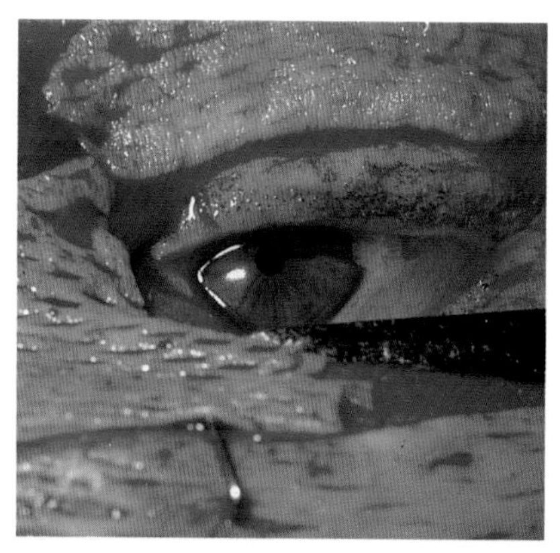

(c) The lids have been incised, retracted and grafted.

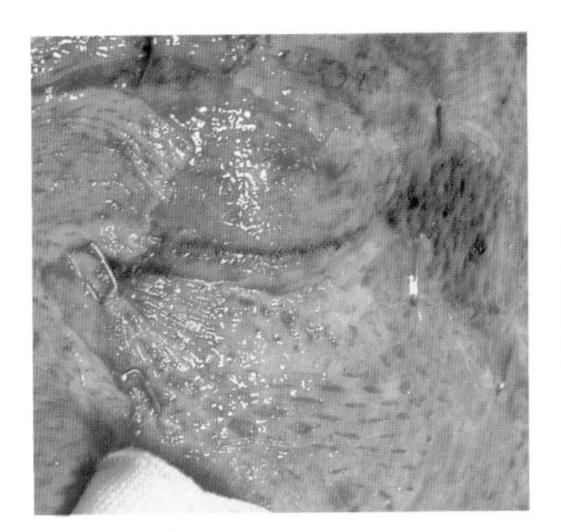

(d) Appearance 72 hours later.

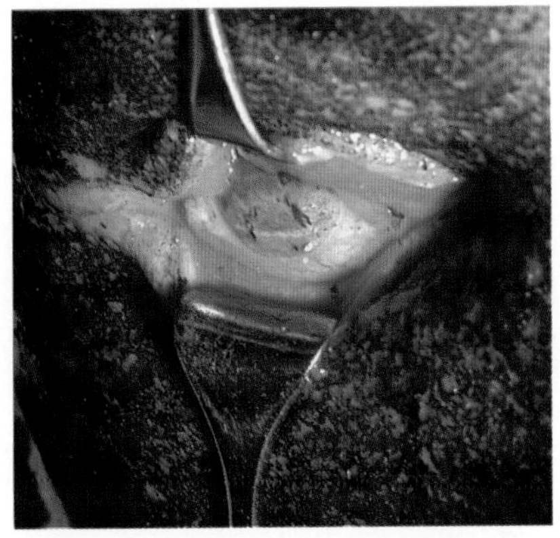

(e) In a separate case, the blink reflex was too slow and deeper damage to the cornea resulted.

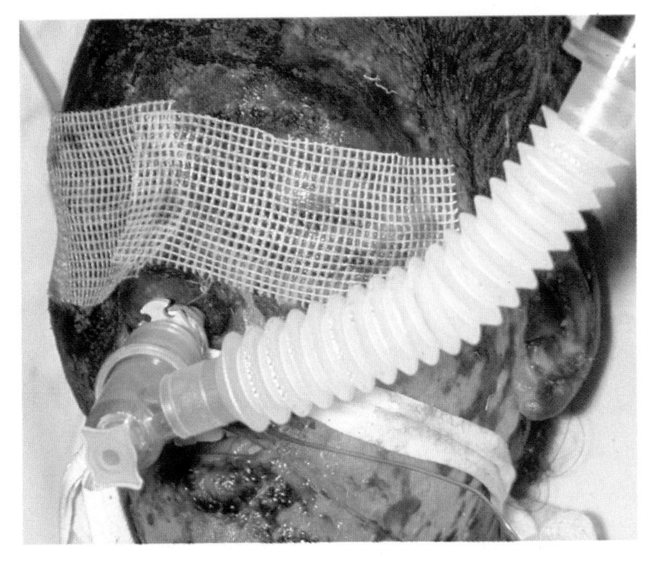

Fig. 5.45 Protection of the cornea during anaesthesia.

corneal cover, and retraction is overcome by generous incision, dissection and skin grafting (Fig. 5.46). An incision is made just above the ciliary margin with dissection of skin from muscle; no tissue has been excised and the defect is made as large as possible. A split skin graft can be held in place with a stent. A tarsorrhaphy must not be performed as the sutures will rapidly tear through the lid, leaving a disfiguring notch. Generous split skin grafting is always required and may need to be repeated.

Eversion of the lower lid is seen more frequently but a rapid response is not as important because the cornea will not be at risk (Fig. 5.47). In this figure an incision is made along the lash margin and a level of dissection developed between the skin and the orbicularis oculis muscles. The lid is advanced as far as possible over the globe to over-correct the deformity, and the defect closed with a graft. Almost no tissue has been excised. The deformity is painful or uncomfortable for the patient and must again be managed by incision, wide dissection and skin grafting. In the lower lid, the graft should be thick or of full thickness in order to provide physical support (Fig. 5.48).

The scalp

The skin of the scalp is well vascularized and rich in hair follicles and should therefore be capable of considerable recovery. After injury, the natural swelling which occurs appears to impair the blood supply, causing infarction in areas of the scalp with a surprisingly large amount shed as slough. In large burns of the scalp, the results of early tangential incision are disappointing; however, this technique is worthwhile in small areas. In cases where the hair has been ignited, full thickness loss may be anticipated but the aponeurosis may remain intact (Fig. 5.49).

Deep burns of the scalp can be very slow to heal where the underlying aponeurosis and periosteum are involved; this frequently occurs in elderly patients (Fig. 5.50). In these cases a conservative approach is

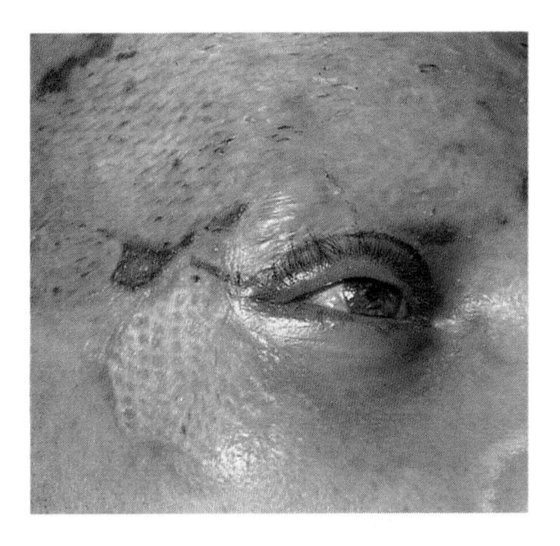

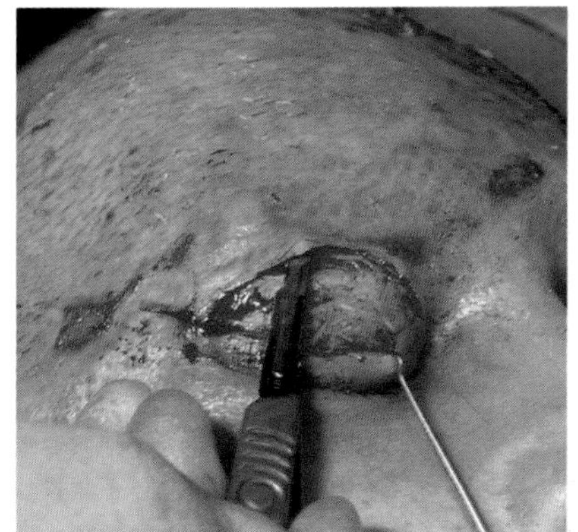

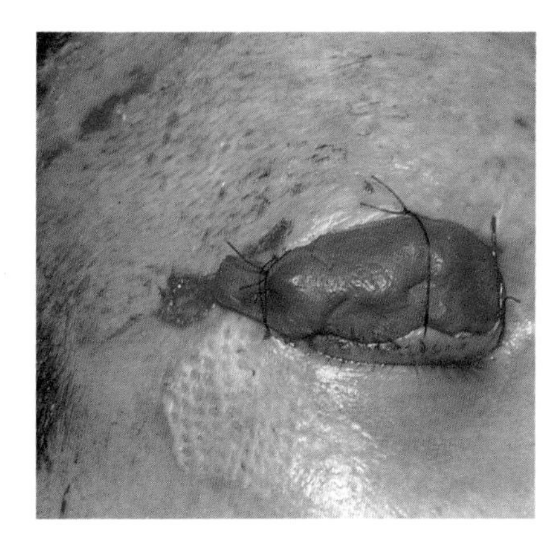

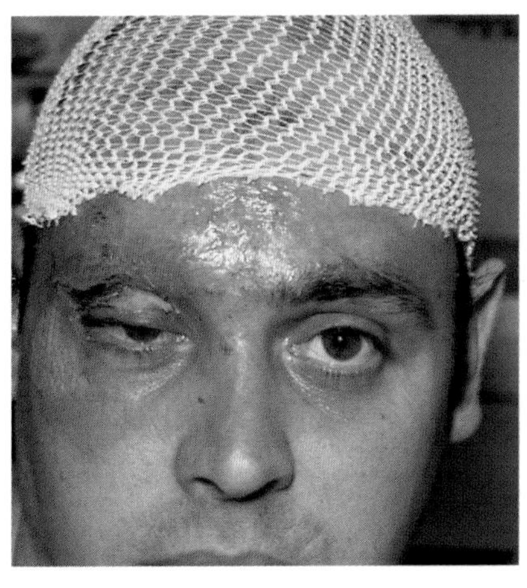

Fig. 5.46 Ectropion of the upper lid. (a) Eversion of the right eyelid. The area to be excised is marked.
(b) An incision of the skin is made.
(c) A split skin graft is held in place.
(d) Appearance one week later.

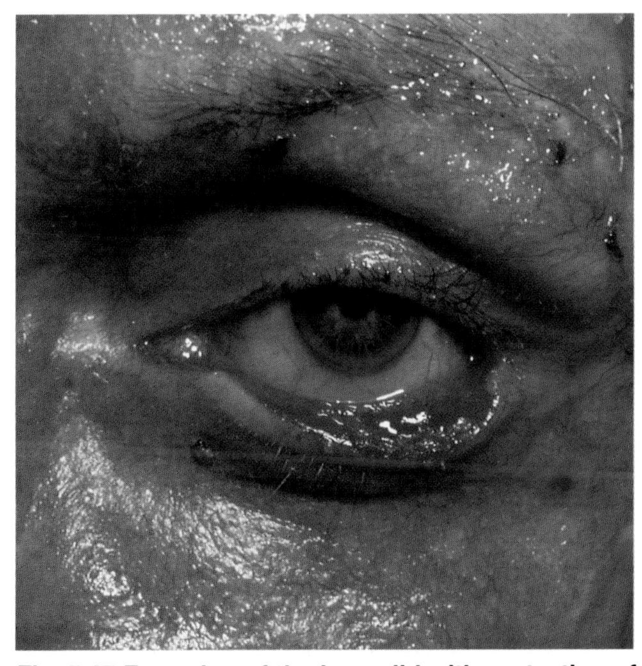

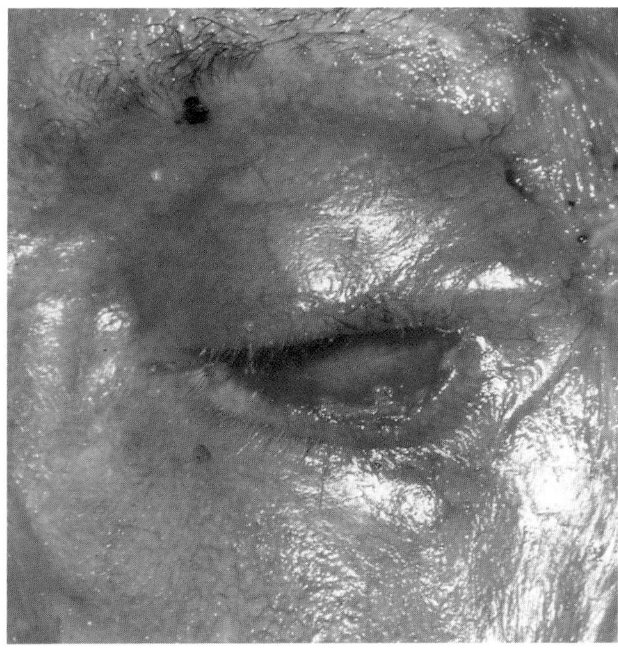

Fig. 5.47 Ectropion of the lower lid with protection of the cornea. (a) The eversion can be seen.
(b) The upper lid remains mobile and there is adequate cover of the cornea.

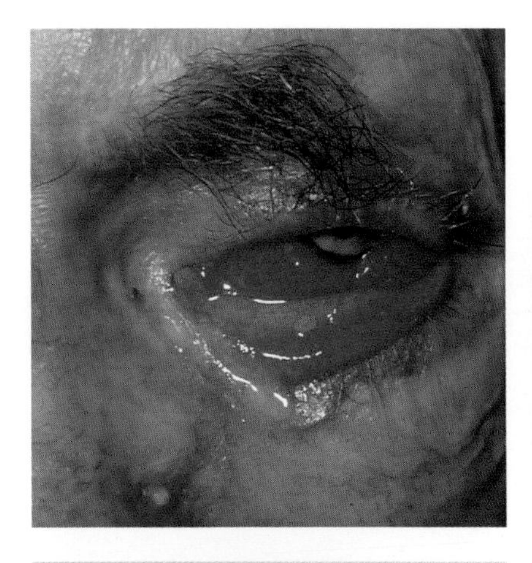

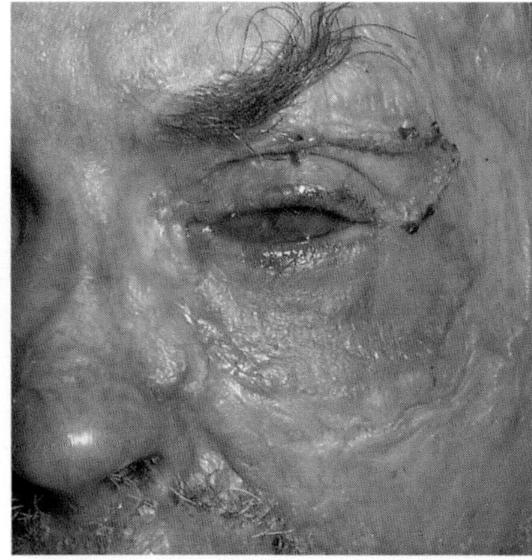

Fig. 5.48 Correction of lower lid ectropion. (a) This is corrected by incision and grafting.
(b) The skin graft must be of full thickness.

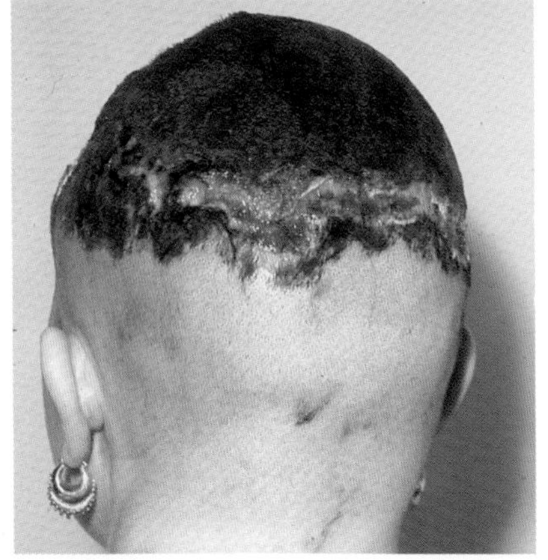

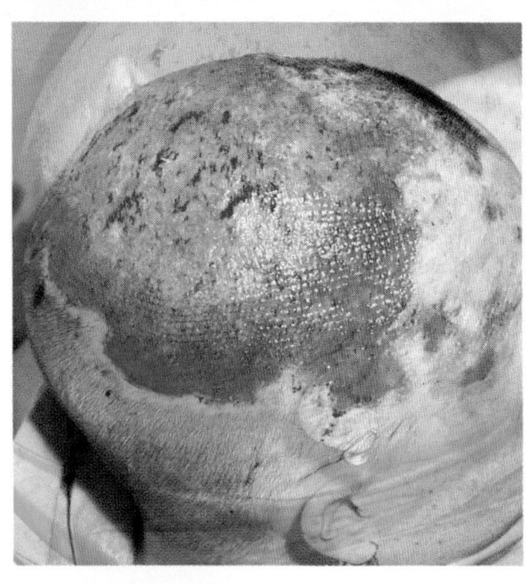

Fig. 5.49 Deep burn of the scalp (a) On presentation. In this case, lacquered hair had been ignited.
(b) Full thickness loss is seen, but the aponeurosis is intact.

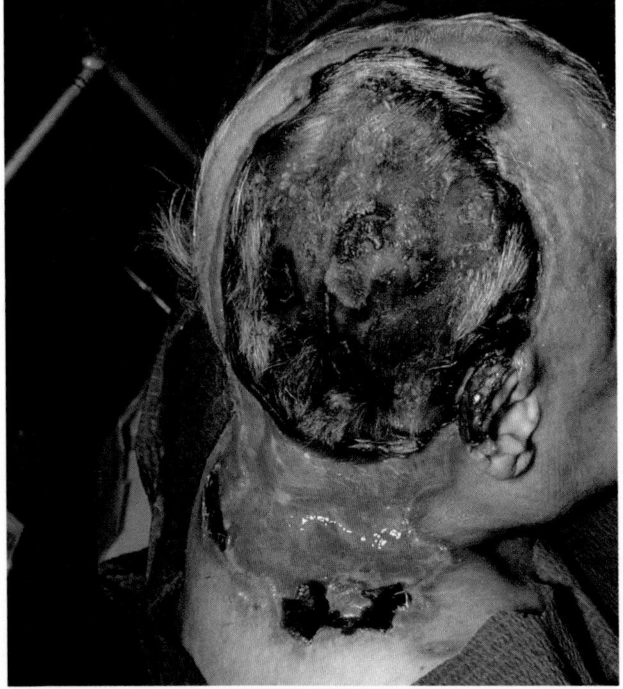

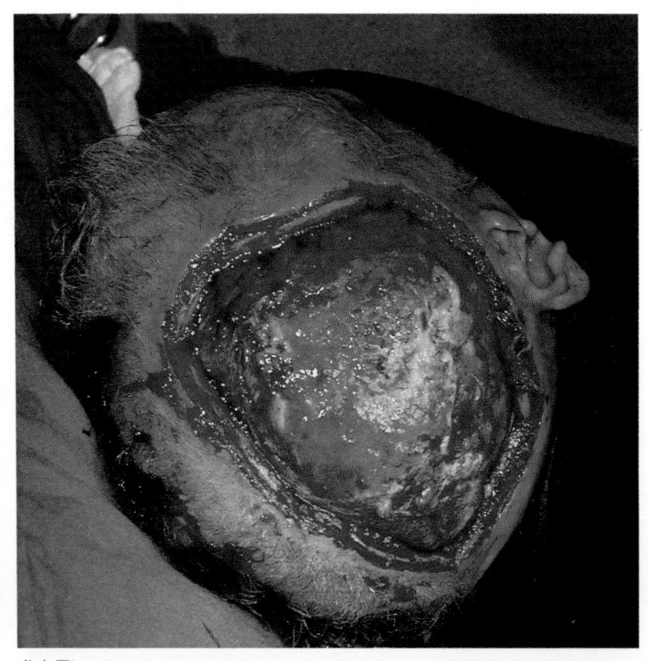

Fig. 5.50 Very deep burn of the scalp. (a) This deep burn involving bone was seen after an elderly woman fell into a fire. The resultant injuries were treated conservatively.
(b) The bone was removed after four months, exposing a healthy layer of granulation tissue which was grafted.

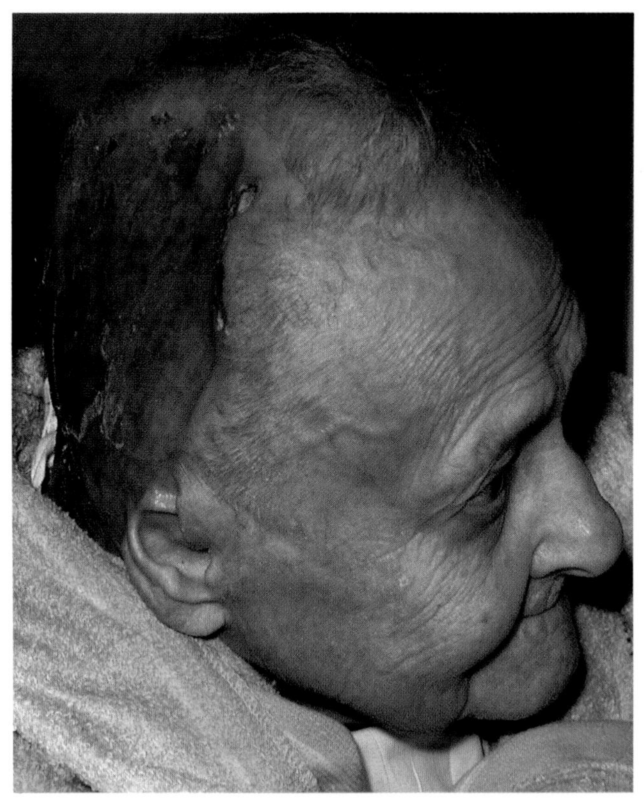

(c) The graft took without delay. There was no evidence of neurological deficit, epilepsy or headaches.

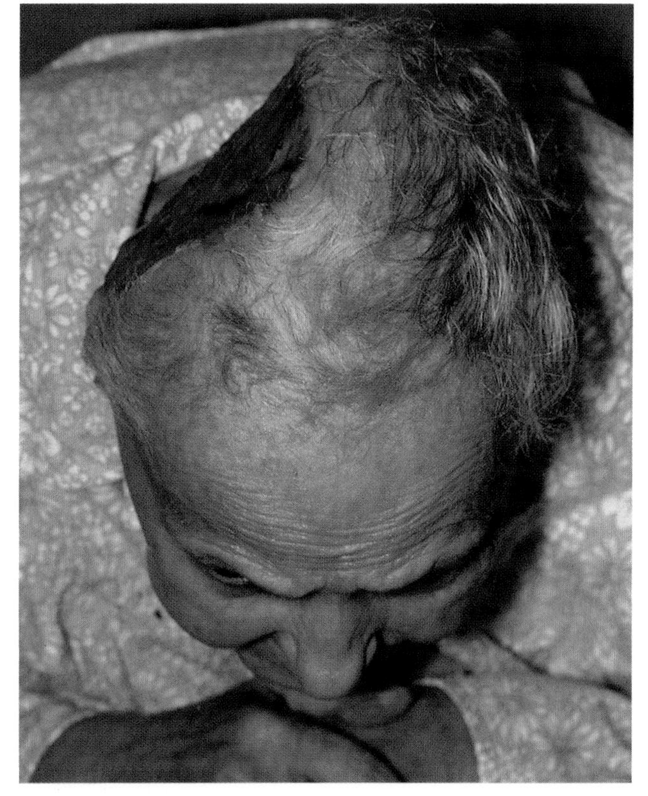

(d) Above view. There is dramatic contour deformity.

often adopted. The exposed bone is cross hatched and removed with an osteotome. Dressings are applied until healthy granulations develop which will permit skin grafting. Occasionally, the inner table of the skull is lost, but grafts can become established on the dura. Small areas of exposed skull can be covered with local skin flaps, and skin grafts applied to the secondary defect.

Perineum and genitalia

Burns to this area are common and tend to be superficial (Fig. 5.51). Although healing is slow and there are obvious sources of local infection, grafting is usually unnecessary. Formal dressings are difficult to apply, thus the area should be kept as clean as possible, with frequent application of a bland antiseptic cream. Swelling of the genitalia is often dramatic, but obstruction to micturition is very rare; therefore, routine catheterization is not essential. However, in very deep burns, catheterization will be necessary (Fig. 5.52).

Burns of the foot

Split thickness skin grafts in this area stand up to wear and tear surprisingly well (Fig. 5.53). Injuries of the foot in diabetic patients may result in amputation (Fig. 5.54), and in cases where the patient is disabled, early excision with skin grafting is recommended (Fig. 5.55).

Bones and joints

Burn injuries deep enough to involve bones and joints are usually seen in patients where consciousness was lost, and tissues were therefore in contact with the thermal agent for a prolonged period (Fig. 5.56). Small areas of bone such as the malleoli, the subcutaneous border of the tibia and the olecranon can be treated with wet dressings. Management of burns involving the joints can often be conservative. Small burns over the elbow or small joints of the foot can be dressed and the patient mobilized. Healing may be slow but will occur with little long term disability. It may be necessary to remove small flakes of dead bone tissue, although granulations will eventually cover the bone completely permitting a skin graft to take. In most cases, durable cover is obtained. Extensive burns involving larger joints such as the shoulder, knee or hip are managed by excision and subsequently covered using a muscle flap to achieve healing more rapidly.

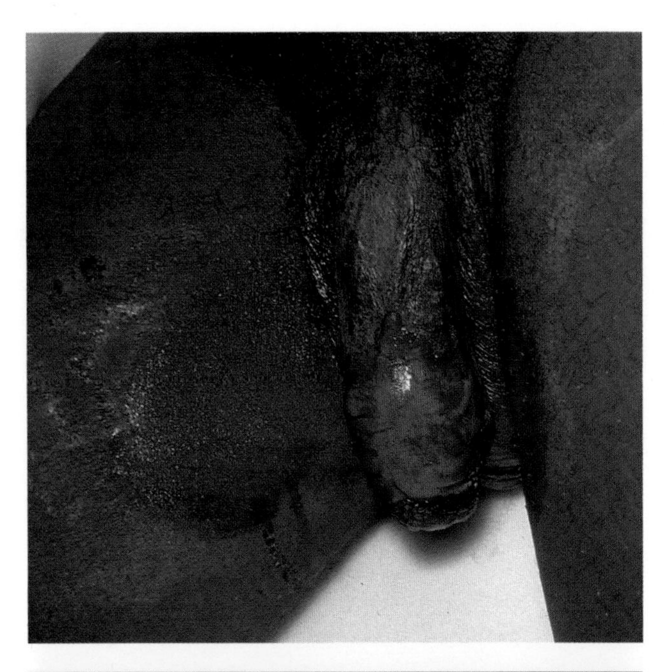

Fig. 5.51 Superficial scald of the penis. Although this is a painful and alarming injury, healing was rapid with no scarring or loss of function.

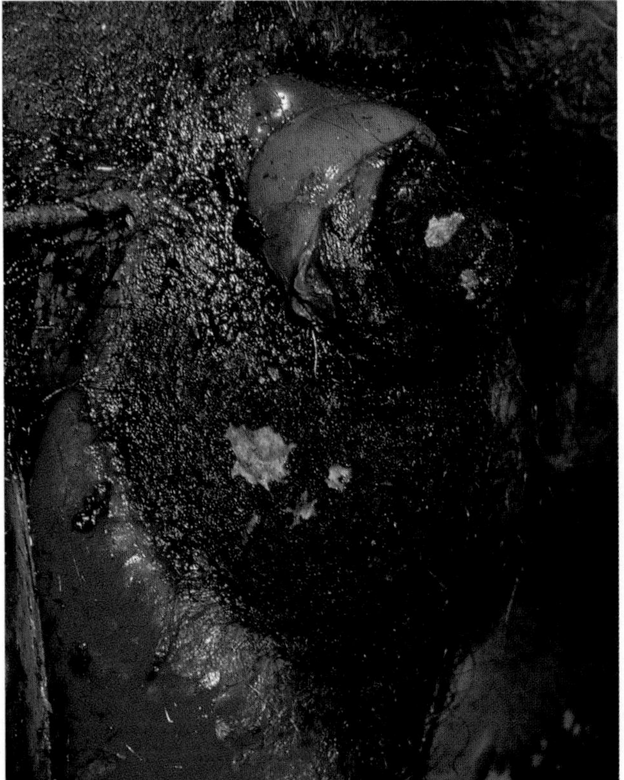

Fig. 5.52 Very deep burn. (a) The tip of the penis has been severely charred.

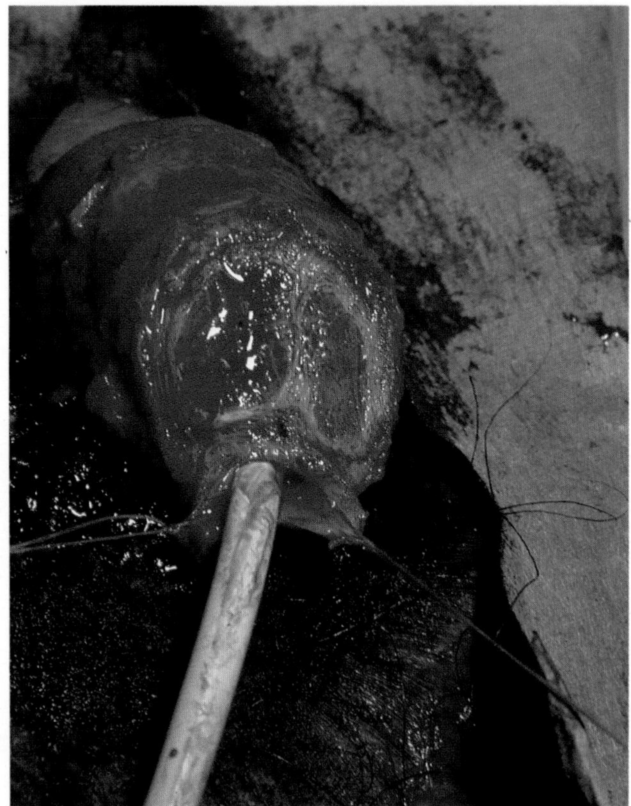

(b) Partial amputation was required to allow catheterization.

occur below the knee, early amputation should be considered. Although slow healing can be obtained, eventual loss of function and ulceration at the ankle may necessitate amputation up to a year after or so following the injury.

PHYSIOTHERAPY

The preservation of movement and function is important throughout all phases of the burn injury. They physiotherapist is very helpful in accepting the

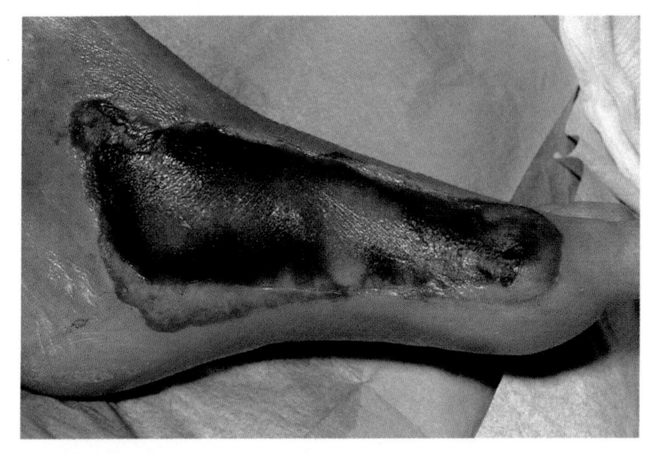

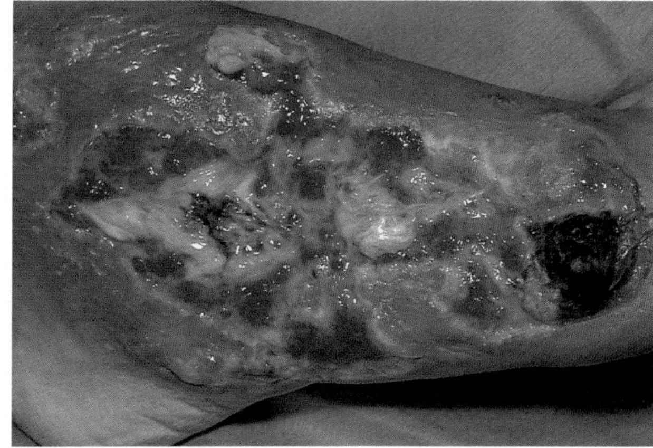

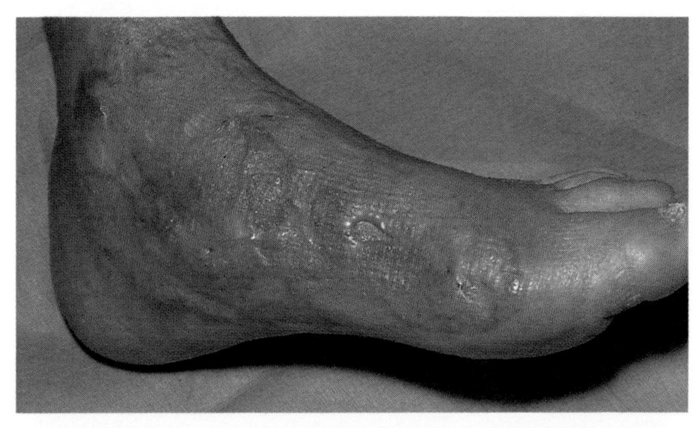

Fig. 5.53 A deep contact burn of the foot. (a) On presentation.
(b) The patient was elderly and declined active intervention. The burn was thus simply dressed.
(c) Activity was continued with the foot treated in a dressing of zinc oxide tape. Healing was complete at six months. No antibiotics or surgery were received, and although numerous small joints of the foot had been exposed, function remained excellent.

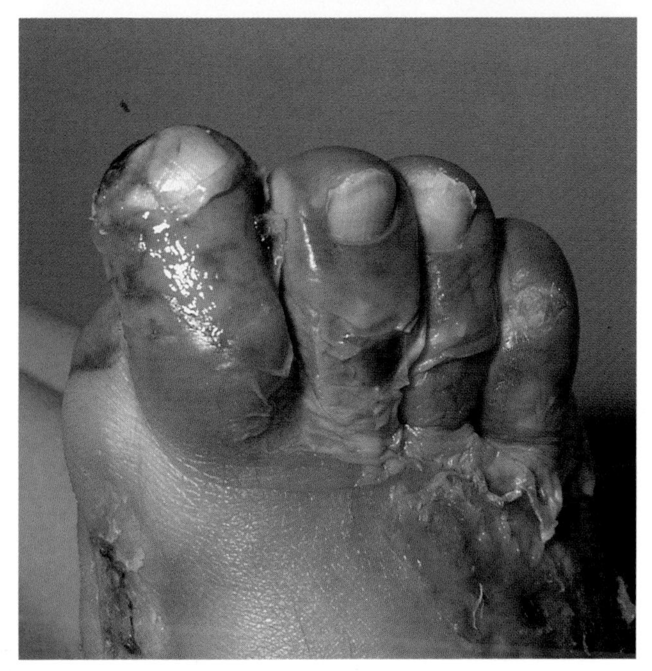

Fig. 5.54 Scald injury of the foot in a diabetic patient. Amputation was necessary.

burden of this responsibility, but other members of the team must be supportive. The needs of each patient must be assessed, and treatment plans outlined and understood by the patient and the nurses. Progress is monitored daily and changes instituted as soon as problems arise.

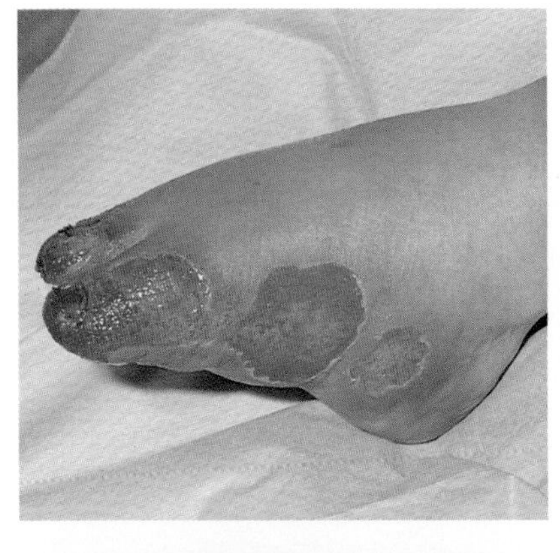

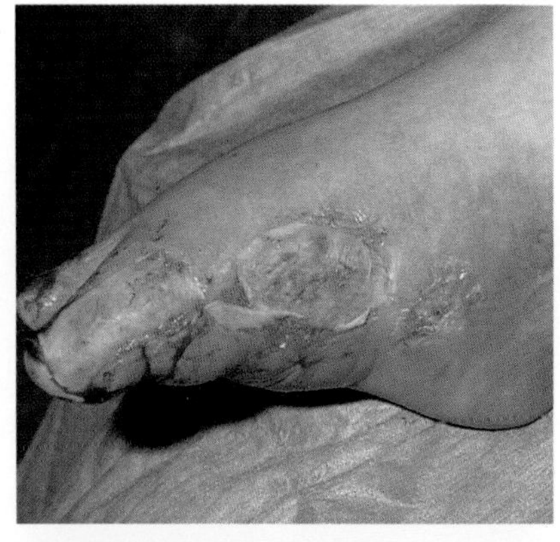

Fig. 5.55 Apparently superficial scald of the foot in a disabled patient. The patient suffered from spina bifida and anaesthetic feet. (a) On presentation.
(b) If treated expectantly, healing is usually very slow.

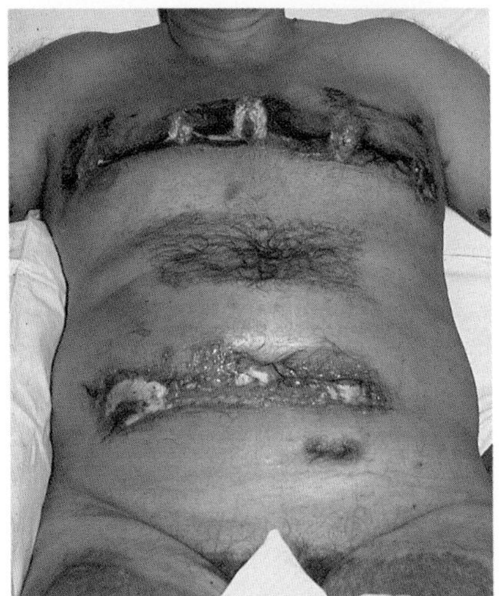

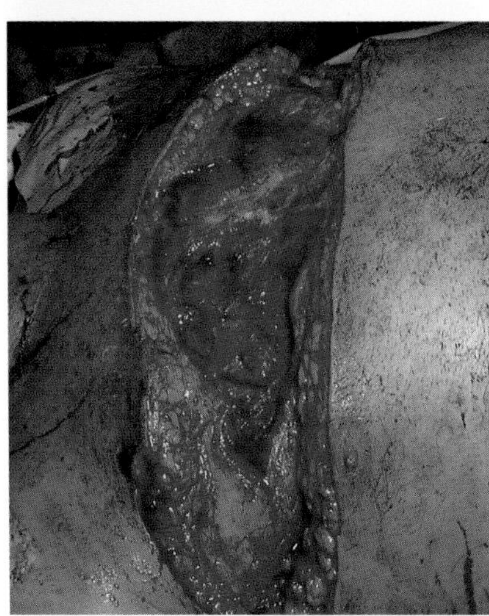

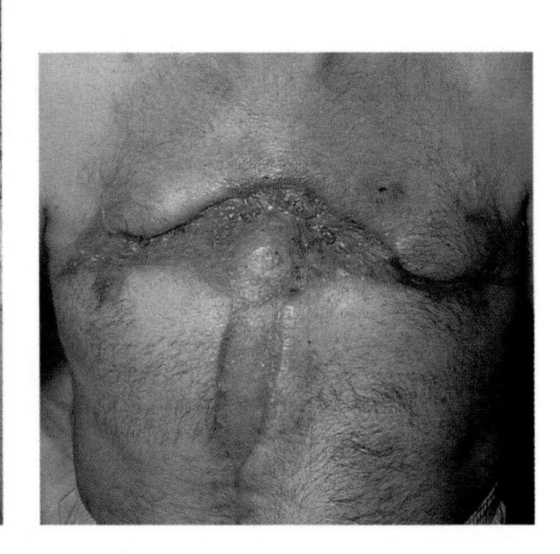

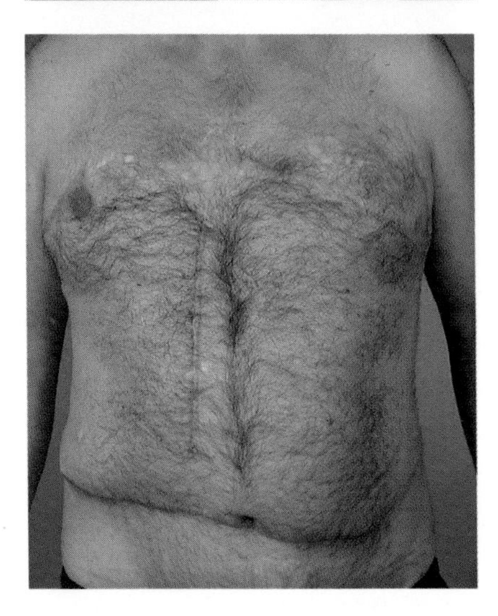

Fig. 5.56 Deep burns, reaching to bone. (a) Three parallel burns were sustained when the patient fell onto electrified rails.
(b) The deepest burns involved sternum, costal cartilages and pericardium, and was closed using an omental flap which was skin grafted.
(c) Appearance two months later.
(d) Eighteen months later, the scars have been excised and narrowed to improve cosmesis.

Early care

Positioning

Burn-injured hands need to be elevated to help gravity overcome oedema and burns involving the trunk and axilla demand that the arms are widely abducted (Fig. 5.57). No pillows should be used in patients with burns of the face and neck, or of the ears. Where the legs are involved they should be elevated until swelling has subsided. Wherever possible, pressure upon the burn should be relieved, i.e. burns of the back are nursed by lying the patient on their front, and burns of the flank are nursed by turning the patient on their other side. If the patient is to be treated by exposure, all raw areas, particularly those involving the flexures, need to be open to the air to

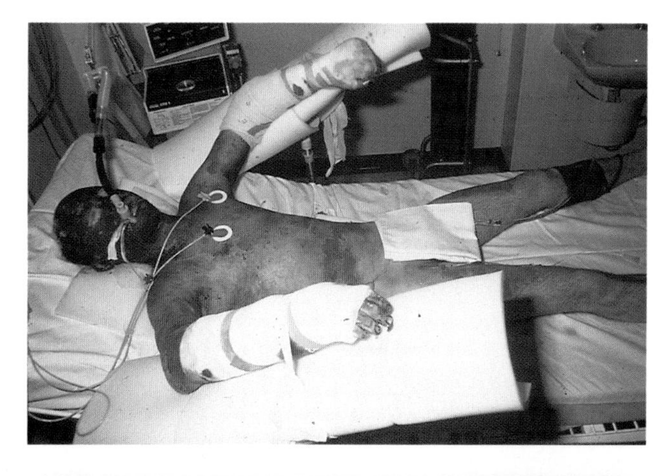

Fig. 5.57 Positioning. The arms are elevated and abducted, the hands will be put through their full range of movement, and splints are made.

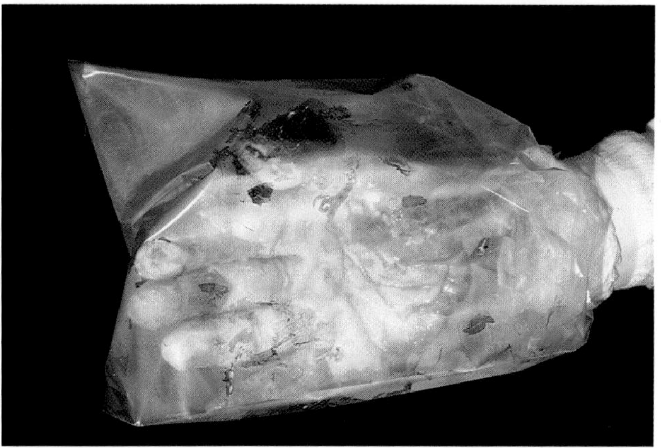

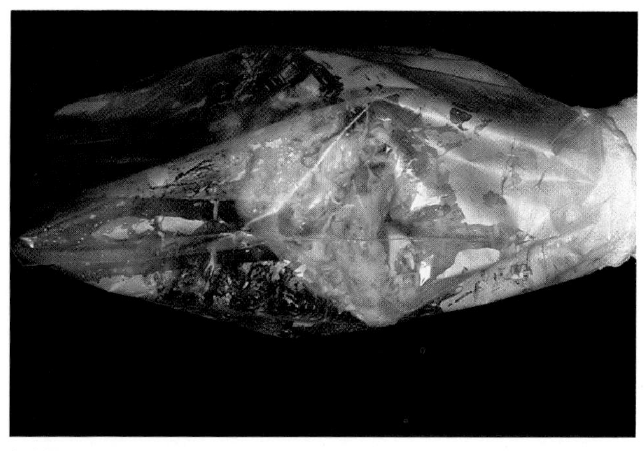

Fig. 5.58 Treatment of hand burns. (a) Full, painless flexion.

(b) Extension of the hand within plastic bags.

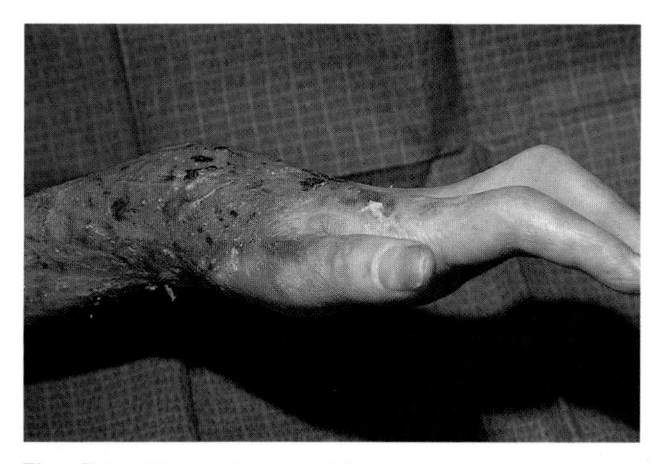

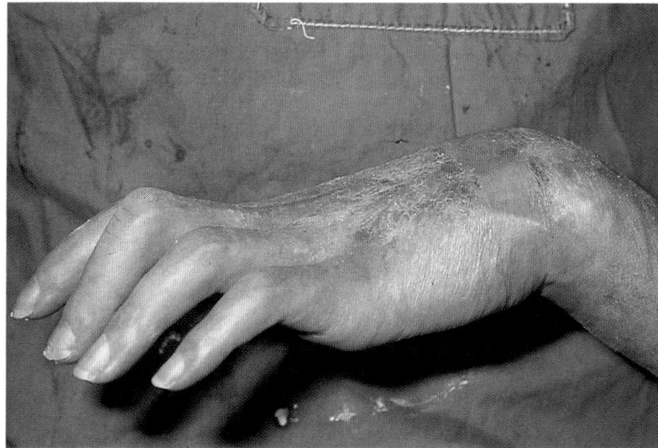

Fig. 5.59 Worsening position of hand over a six month period. Constant splinting has been applied but has failed to correct flexion at the wrist. (a) On splinting.

(b) Six months later, ligaments and muscles have shortened permanently.

encourage rapid drying.

Whenever possible, limbs must be put through the full range of movement, both actively and passively, and this is easiest to achieve at the time of dressing change and when the patient is in the bath. The presence of a physiotherapist during dressing changes is to be thoroughly recommended.

Patients with burns of the hand require special care. When the burns are treated in hand bags the patient find movements fairly easy and painless (Fig. 5.58), but they must be encouraged to try and regain the full range of movement as quickly as possible. Hand burns treated in dressings do not show the same rate of improvement. The hands will be oedematous for longer, the fingers stiffer and the hands will tend to fall into a position of wrist flexion, metacarpophalangeal hyperextension and flexion of the fingers. This will rapidly worsen and become more difficult to overcome, with shortening of both small muscles and joint ligaments. Significant burns of the hand treated in dressings need to be splinted for most of the time (except for meals and periods of exercise) to ensure that the wrist is dorsiflexed, the metacarpophlangeal joints flexed to 90°, and the fingers in full extension. The splint must be well padded to avoid pressure, particularly over the proximal interphalangeal joints, and easy to apply and remove.

Development of a dry leathery eschar following exposure effectively splints and limits movement pressures similar to splinting. If these are forced the eschar will crack and, apart from causing pain, will allow the ingress of bacteria. Passive movements must therefore be controlled. Inhalational injuries will obviously add an extra dimension to the care the burnt patient requires. When paralysed and receiving artificial ventilation, muscles waste very rapidly, but the patient will not complain when limbs are put through their full range of movement. Hands need to be splinted, arms abducted and elevated and, when possible, the legs should be elevated also. Pulmonary function is maintained using the standard intensive care technique, and tipping and chest percussion should not necessarily be avoided because the patient is suffering from burn injuries.

Post grafting

Shearing forces over freshly grafted areas should obviously be avoided, and where grafts have been covered with bulky dressings, active and passive movements should be restrained. Hopefully the surgeon will have applied splints to grafted hands (Fig. 5.59) and across grafted joints to maintain the best possible position (Fig. 5.60). Limbs with exposed grafts may be moved gently in the immediate postoperative phase, but otherwise activity must be delayed until the time of the first dressing. The physiotherapist should always try to be present at this time, even in the operating theatre, to flex and extend joints and supervise the manufacture of thermoplastic splints.

No graft will ever take totally and most will need some form of dressing for two or three weeks before the graft is ready to withstand the prolonged pressure that will help to prevent their contraction. The bed beneath the graft will contract, and the more granulating tissue, the more rapid and severe the force of contraction will be. It can be overcome in part by pressure, which will be more effective the sooner it can be applied. However, when applied too early, areas of underlying graft will be lost, resulting in the growth of more granulating tissue and thus more contracture. Small patches of graft loss should be treated with frequent dressings of 1% hydrocortisone ointment or zinc oxide tape, and the splint reapplied. In larger areas granulations are shaved down, fresh grafts applied and covered with a soft and bulky splint.

As soon as grafts have established themselves they should be supported with elastic pressure or thermoplastic splints to retain suppleness and to minimize the tendency to contracture and hypertrophic scarring. Hands which have been skin grafted should be placed in elastic gloves as quickly as possible.

Overall mobility must be encouraged and burnt limbs covered with supportive dressings and overlying elastic bandages. An exercise programme and regular visits to the physiotherapy department and occupational therapy workshop should be started as soon as medical conditions permit.

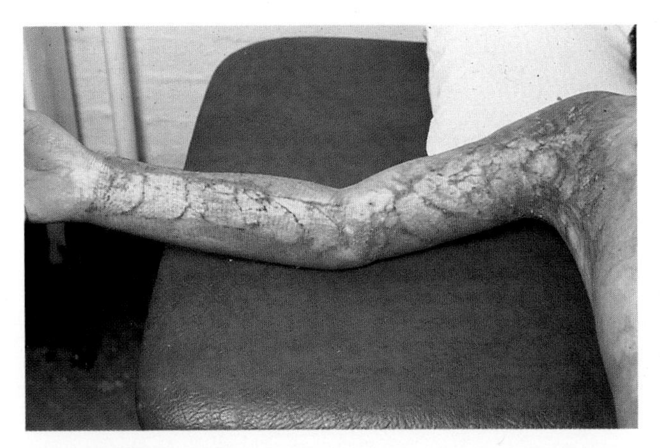

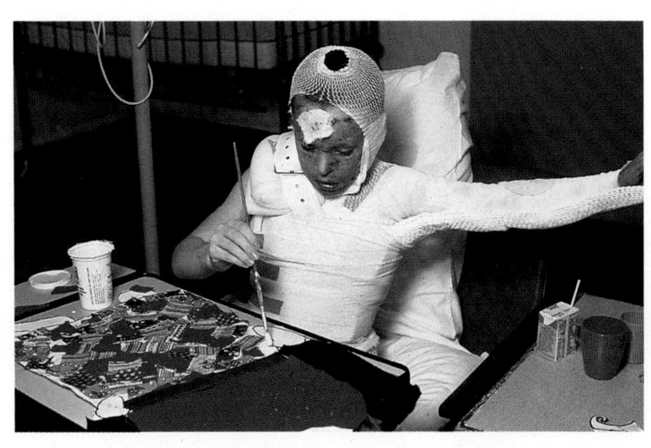

Fig. 5.60 Skin grafts across joints. (a) Prolonged splinting is required. (b) The axilla is very difficult to control.

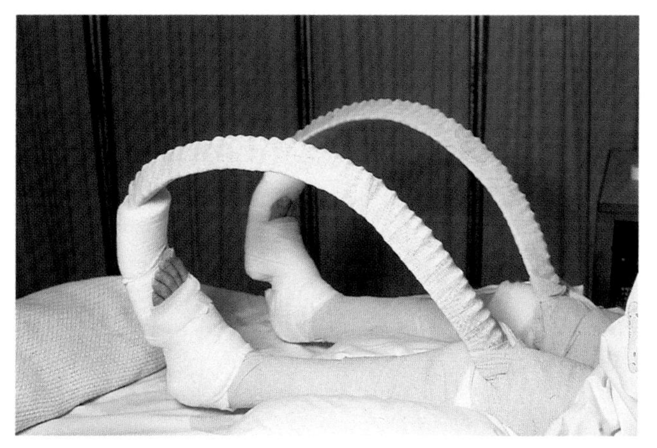

Fig. 5.61 Splinting to achieve dorsiflexion at the ankles. No undue pressure is placed on the dorsum of the feet or the calves, which were deeply burned.

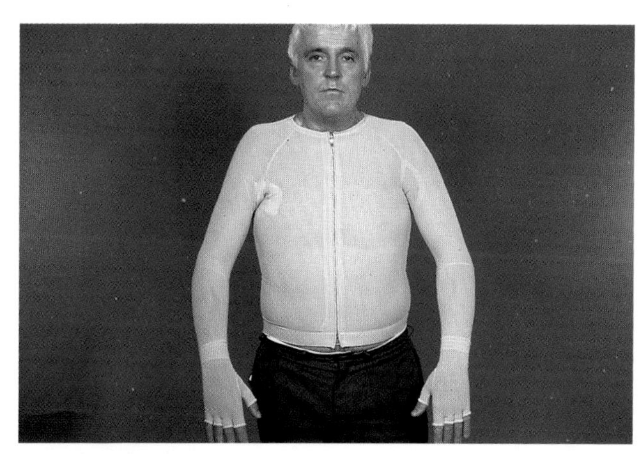

Fig. 5.62 Elastic pressure garments.

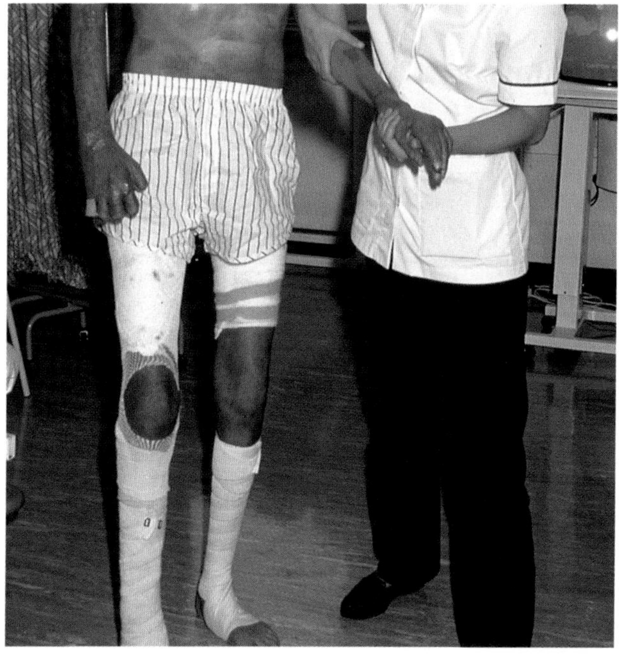

Fig. 5.63 Exercise program. (a) Muscle strength is increased and the patient is encouraged to pull against the forces of scar contracture.
(b) In this patient the grafted lower legs are supported with a firm elastic bandage.

Late treatment and rehabilitation

When the grafts have largely healed, treatment is directed to regaining as much function as possible. The contracting burn scars will tend to decrease function and limit the range of movement. These must be resisted by building muscle strength and confidence, and by the early and effective application of splints (Fig. 5.61) and pressure garments. Scars extending across joints should be splinted with a thermoplastic splint pressing directly onto the scar, and worn continually. it is only removed for meals, dressings, baths and periods of exercise. Areas of scar or skin graft not across joints are covered with a tubular elastic bandage or a custom-made elastic pressure garment (Fig. 5.62), to be worn at all times as soon as the wounds permit.

Splints need to be regularly remoulded to ensure that they are constantly working against the pull of the scar, and that the advantage gained is maintained and steadily improved upon. At the same time the patient must be actively and enthusiastically involved in an exercise program which will continue to increase muscle strength (Fig. 5.63).

Long-term hand function is directly related to the time the burns take to heal, and in most centres early surgery is directed towards this end. Hands that were grafted and subsequently heal within two weeks usually do well. Long-term splinting is unnecessary although elastic gloves should be worn in all

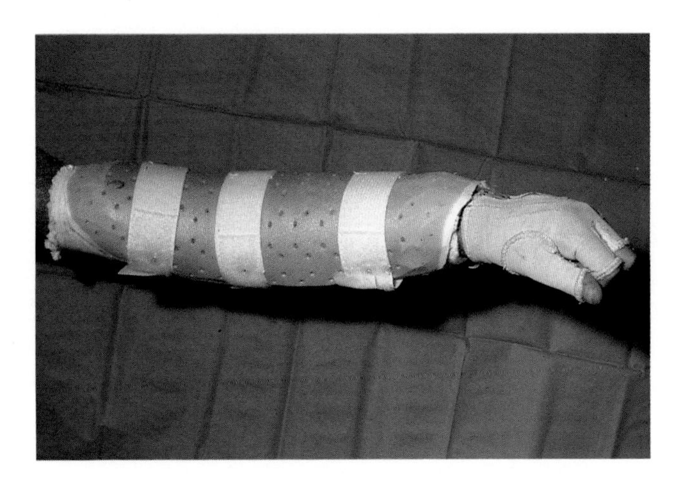

Fig. 5.64 Night splint worn over the pressure garment. This is easily applied and removed, and is held firm with velcro straps.

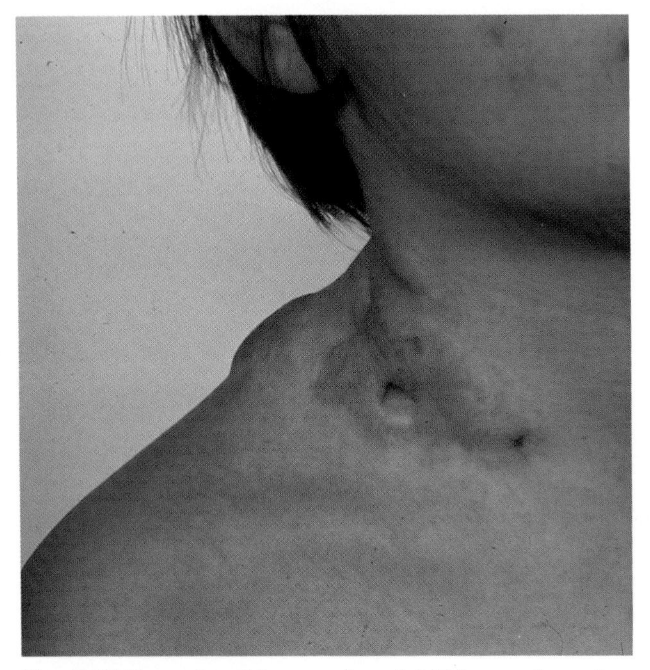

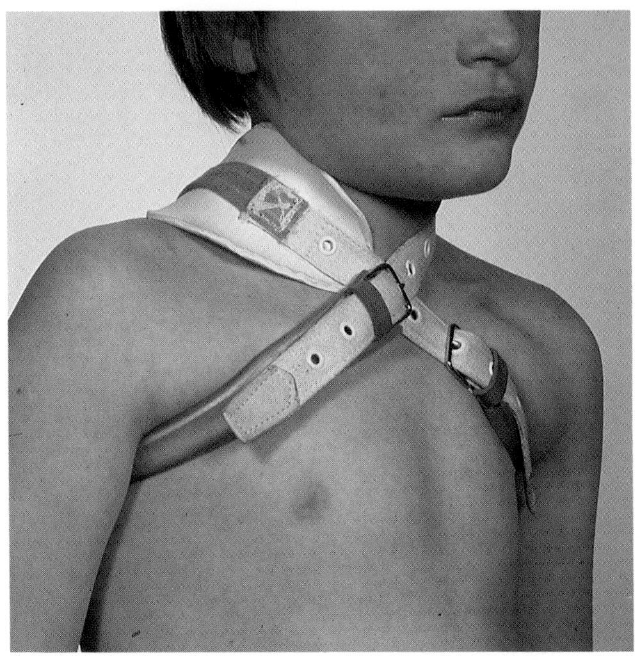

Fig. 5.65 Hypertrophic scarring. (a) This very hypertrophic scar causes irritation.

(b) Local pressure is applied to decrease itchiness.

cases of hand burns. Where healing has taken longer some loss of function in the short and medium term is inevitable, and treatment is directed towards active and passive exercises and stretching, night splints (Fig. 5.64) and as much occupational therapy as possible.

All splints must be worn until the forces of contracture have weakened. These forces can be overcome by active stretching and muscle strength; on occasions this may be for three or four months. Pressure garments nearly always need to be worn for a much longer period. They are helpful in relieving the intense itching and irritation that is such a characteristic feature of hypertrophic scars, and they undoubtedly modulate the surface of the scar and speed maturation (Fig. 5.65). Silicone gel sheeting can sometimes flatten scars quickly. It does not need to be applied beneath a pressure garment. Concave areas of the body can be filled with pads of elastomer which help to transmit pressure onto these difficult areas (Fig. 5.66).

Garments in young children need to be refitted every two to three months, and worn for 15 to 18 months until the scars have become pale.

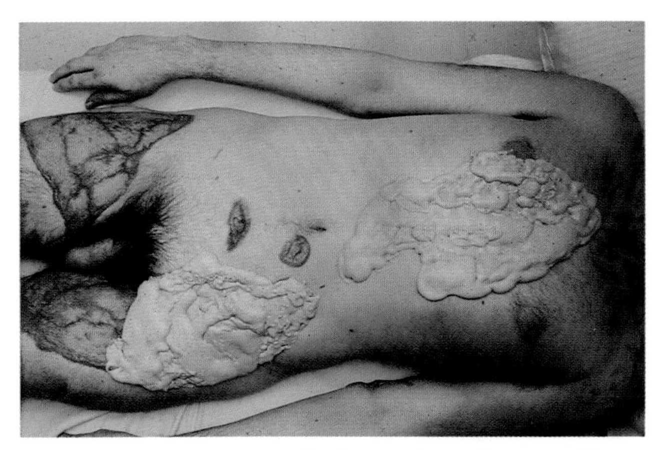

Fig. 5.66 Elastomer applied to awkward areas. These are held in place with adhesive tapes and a pressure garment is worn over them.

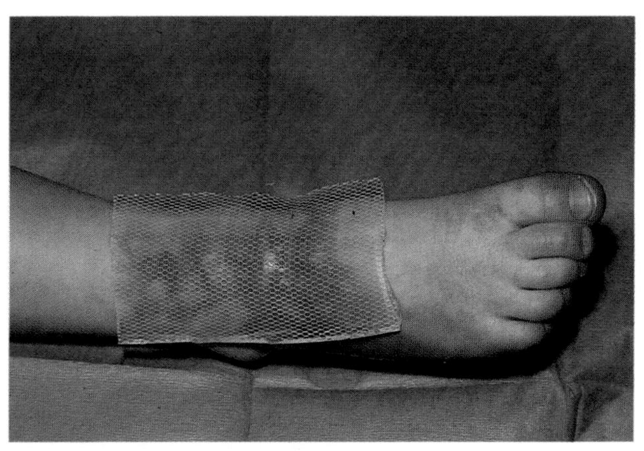

Fig. 5.67 Silicone gel sheeting.

6 Sequelae

SCARS

Burn wounds which take longer than two weeks to heal will result in scarring; generally, the time taken for the wound to heal is directly related to the severity of the scar formed. Some scars may mature and soften rapidly, but in the majority of cases some degree of hypertrophic scarring occurs, which takes from 18 months to two years to mature. Burns which have been covered with meshed skin grafts show a typical reticular pattern (Fig. 6.1). The development of a hypertrophic scar can be seen in Fig. 6.2. The scarring is more severe at the wound edges where dermal damage and repair is maximal. Bad hypertrophic scars are frequently seen in epileptics, particularly in those who are taking phenytoin.

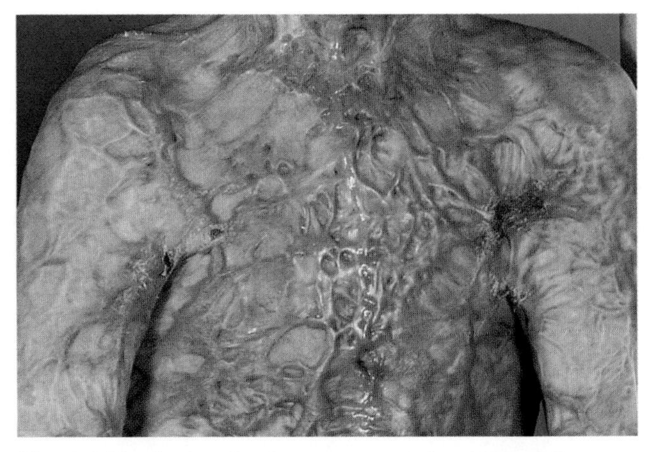

Fig. 6.1 Typical reticular pattern of hypertrophic scarring.

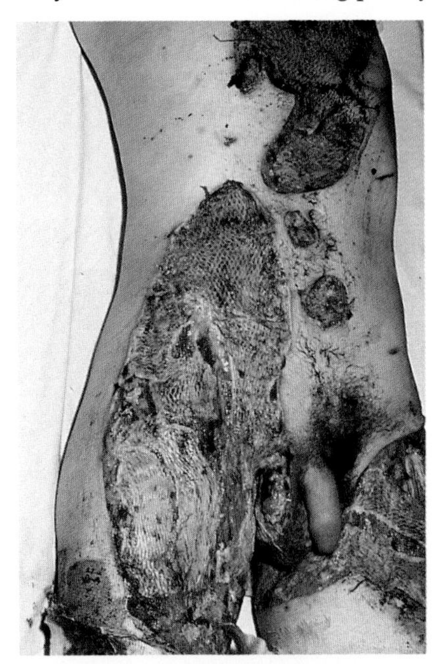

Fig. 6.2 Development of hypertrophic scar in an epileptic patient. (a) The sites have been grafted. Gradual development of scar tissue is seen.

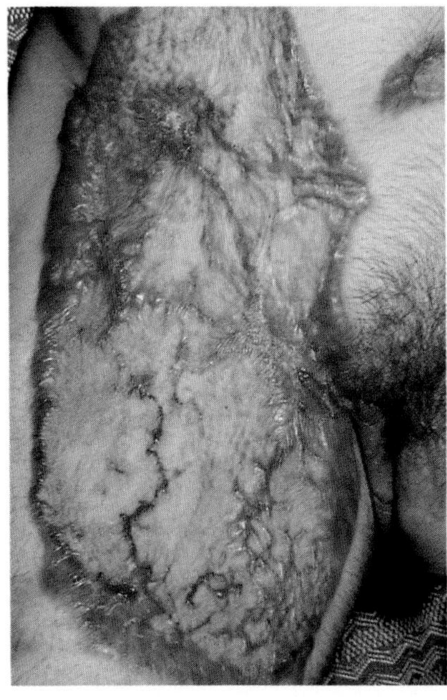

(b) Three months later.

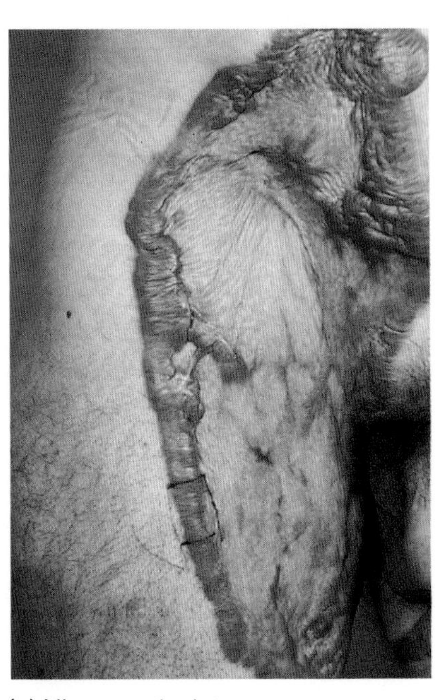

(c) Nine months later.

Initially the scars are red, raised and itchy. As the scar tissue contracts, the subsequent pull on adjacent tissues may limit the range of movement or cause deformity. The thickness of the scar increases for approximately three months and then remains constant for around one year (Fig. 6.3). This then becomes paler and softer (Fig. 6.4); when fully mature, it turns white, atrophic and inelastic, with an uneven contour (Fig. 6.5). Histologically, an increase in collagen, elastin and fibroblasts can be seen in this type of tissue (Fig. 6.6). The scar remains vulnerable to trauma and is relatively slow to heal. In children, inelastic bands of scar tissue may limit normal growth. In these cases, the bands should be divided to overcome deformity and allow the introduction of additional skin.

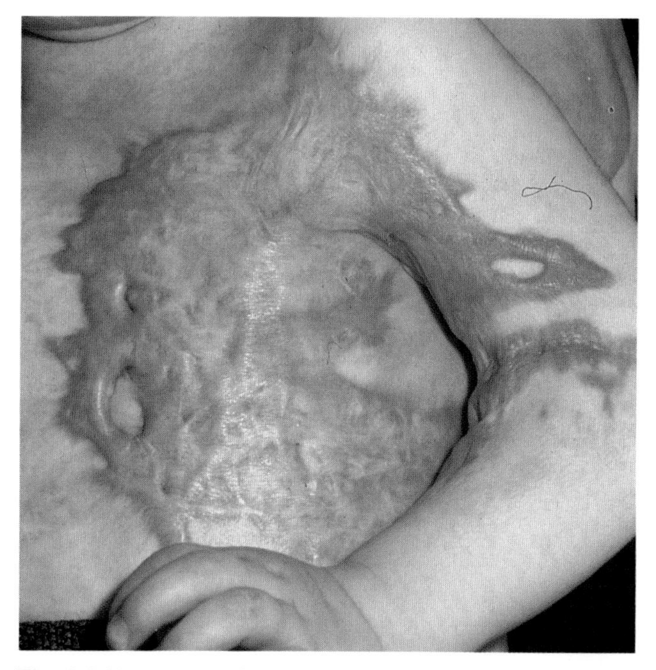

Fig. 6.3 Hypertrophic scarring following spontaneous healing. This injury was due to scalding.

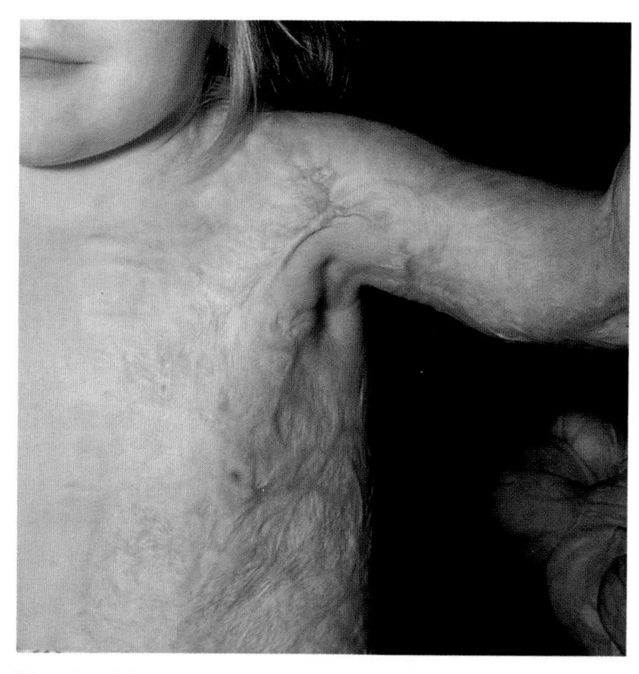

Fig. 6.4 Mature hypertrophic scar. This has matured and become soft and supple but some distortion of breast development can be anticipated.

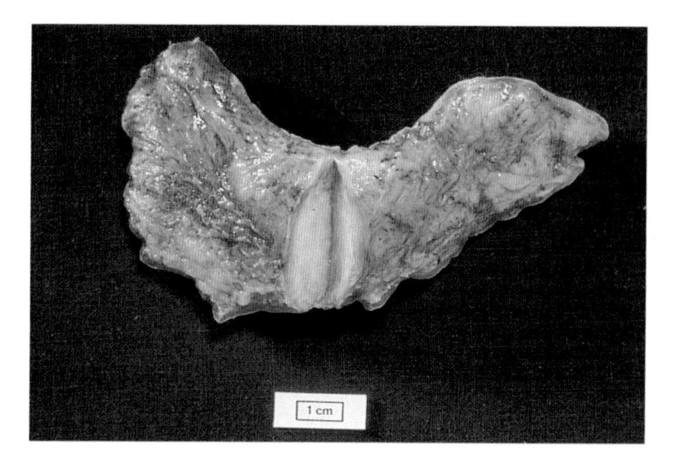

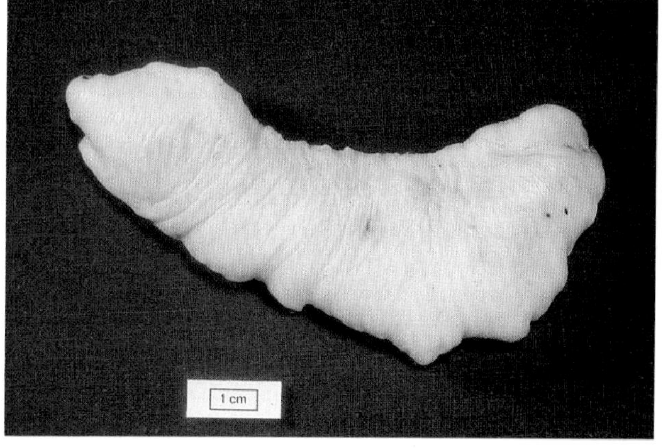

Fig. 6.5 An excised plaque of scar tissue. (a) Deep surface. (b) Obverse view.

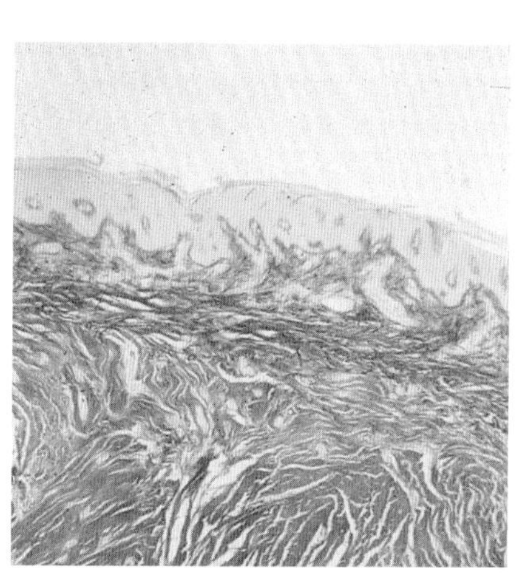

Fig. 6.6 Histology.
(a) Thick section.
(b) Dense scar tissue replaces dermis.

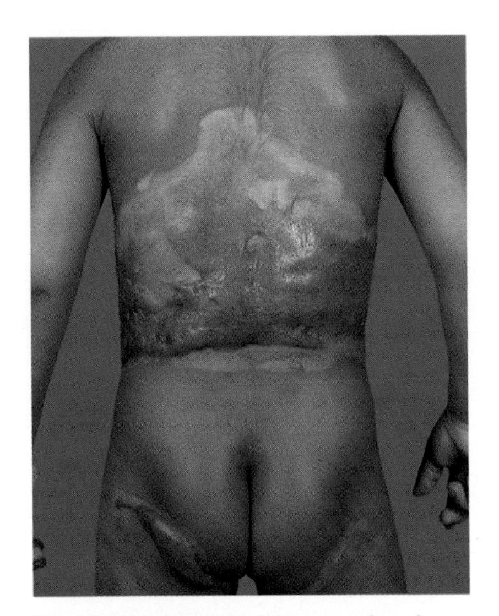

Fig. 6.7 Hypertrophic scarring in pigmented skin.

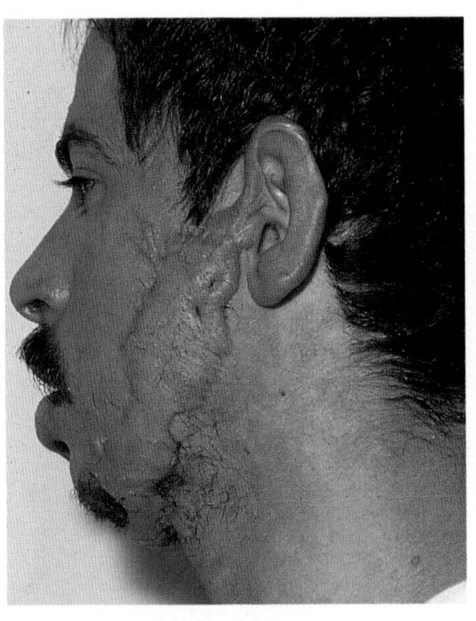

Fig. 6.8 Scar infection. (a) Buried hairs and chronic local infection have arrested resolution of this scar. Excision will be necessary.

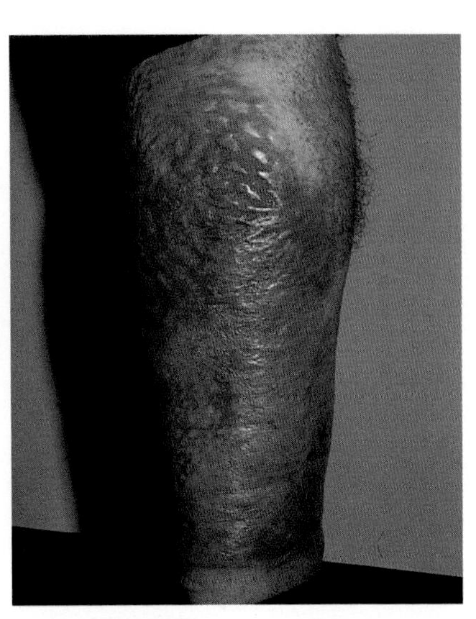

(b) A streptococcal cellulitis has been confined to an area of hypertrophic scar.

In the majority of patients, scarring is hypertrophic rather than kelodial. Hypertrophic scarring does not spread to uninjured tissues and eventually matures and resolves. However, genuine keloid scars gradually increase in size and do not improve. Management is directed towards the control of symptoms and prevention of contraction. The intense irritation of this type of tissue is difficult to control, particularly in children. Medication with antihistamines may help at night, but tight fitting elastic pressure garments afford the most relief. The exact mechanism of suppression of itching is not known, but constantly applied pressure controls this response and also promotes flattening of the scar tissue. Fitted elastic garments are worn as soon as healing permits. Children are remeasured, with new garments fitted every two months; in adults, garments are replaced every four months. Areas that fail to respond well to garments may benefit from direct contact from a silicone gel sheet, or a mould of silicone elastomer. The scarring tends to be itchier, thicker and takes much longer to resolve in those with pigmented skin (Fig. 6.7).

In burn injuries of the hand, the wearing of elastic gloves is generally well tolerated and results are encouraging. Together with the obvious cosmetic advantage, itching is controlled and function is partially restored. The development of webbing between the fingers can be problematic but this often responds to a soft plastic foam pad worn beneath the glove.

Scars which run across joints will contract, causing deformity. This must be resisted with splinting. Areas which are likely to contract such as the neck, axilla, elbow and the hand should be splinted as soon as healing will permit. Skin grafts applied over bony prominences will break down if pressure from a splint is applied too early; if applied too late, the forces of an established contracture are too great to be overcome.

Areas across joints must take priority for grafting, to enable healing and subsequent splinting to take place promptly. When control is achieved, splints can be removed for about one hour during mealtimes, for physiotherapy and for dressing changes. At all other times the splint must be firmly and diligently reapplied. Thermoplastic materials are the most useful: they are quick to remould, easy to clean and have a surface modulating effect when applied without a dressing.

Complications which may inhibit resolution can arise with scars; these include infection (Fig. 6.8), development of abscesses due to buried epidermal appendages (Fig. 6.9), or atrophy of the scar tissue (Fig. 6.10).

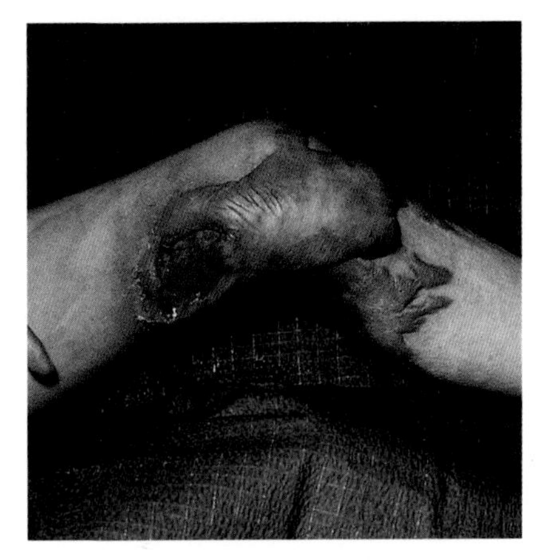

Fig. 6.9 Abscess development. This abscess developed spontaneously in a scar which was nearly two years old.

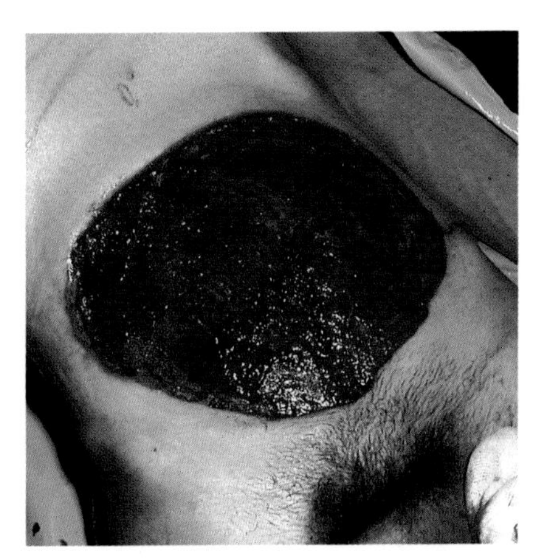

Fig. 6.10 Atrophy of scar tissue. (a) A full thickness wound of the abdominal wall has been excised and is granulating.

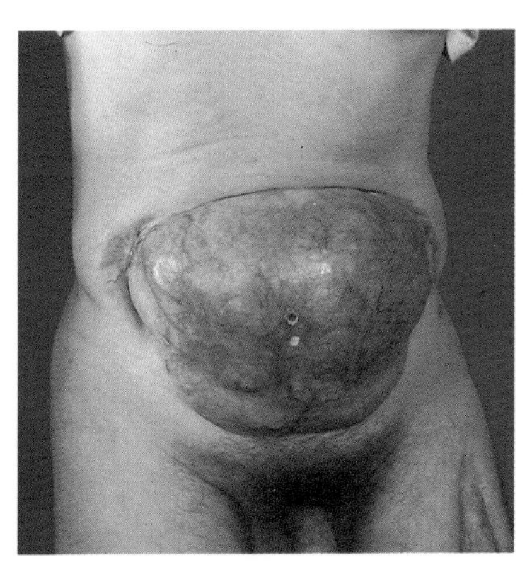

(b) After grafting the scar tissue has become atrophic and cannot adequately contain the viscera.

SPLINTS

Serial splinting can overcome some contractures, particularly at the elbow, hand and knee, but are of little use at the neck or axilla (Fig. 6.11). In the latter cases, the scar must be surgically divided and additional tissue introduced in order to correct the deformity. Alternatively, the neck may be splinted with rings of plastic sucker tubing, joined with a small piece of wooden dowell (Fig. 6.12). These rings can be added as the forces of contracture are overcome, and are inexpensive and re-usable. They are also well tolerated by the patient. When the graft has taken and is well established a thermoplastic splint can be accurately moulded to maintain pressure over the graft and maintain extension of the neck (Fig. 6.13).

In children, burn injuries of the hand are particularly difficult to splint but the oyster shell principle can be successful (Fig. 6.14). An impression of the dorsal and palmar surface of the hand is taken at the time of stitch removal and a plaster model made. Rigid acrylic shells are cast from this model, and the front and back are bolted together encasing the hand in the 'sandwich'. This type of splint is well tolerated and can be worn 24 hours a day for several months. In a child, physiotherapy will not be necessary to mobilize the hand again; in an adult this immobilization would result in permanent stiffness.

Application of pressure to the face can be particularly difficult. The wearing of an elastic face mask, the traditional Jobst mask, is less well-tolerated: the patients feel they have a sinister appearance and the velcro attachments easily become entangled in the hair (Fig. 6.15) but they are ideal for applying pressure to the chin. Alternatively, a transparent flexible plastic mask can be worn, which is built on a cast of the patient's face (Fig. 6.16) and is generally well tolerated. The masks are removed only at mealtimes and can be made to fit specific areas of the face. Their use minimizes the need for secondary reconstruction. The amount of pressure applied can be adjusted by building up or grinding down areas of the original plaster model before remaking the mask.

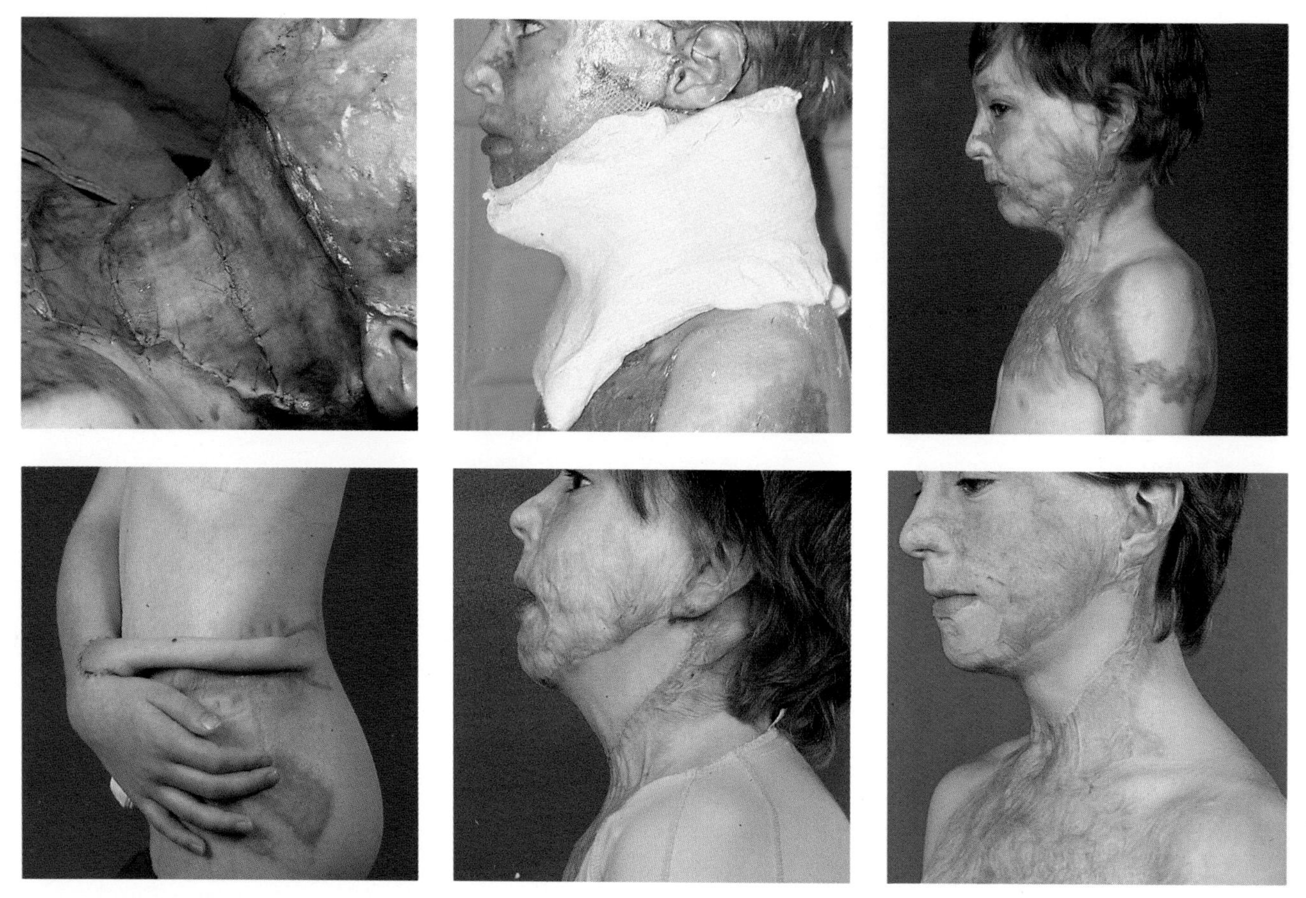

Fig. 6.11 Splinting of the neck. (a) A contracture of the neck was very widely released and grafted.
(b) A splint of plaster of paris was applied. The splint should apply pressure to the graft and maintain extension of the neck, but those applied whilst the patient is anaesthetized rarely do both.
(c) The graft was patchy and the condition recurred.
(d) A tubed pedicle is raised and carried on the wrist.
(e) It may then be inset into the neck. The scar tissue can be completely excised and a full thickness tissue flap inserted.
(f) The contracture will not recur and the cosmetic result is excellent.

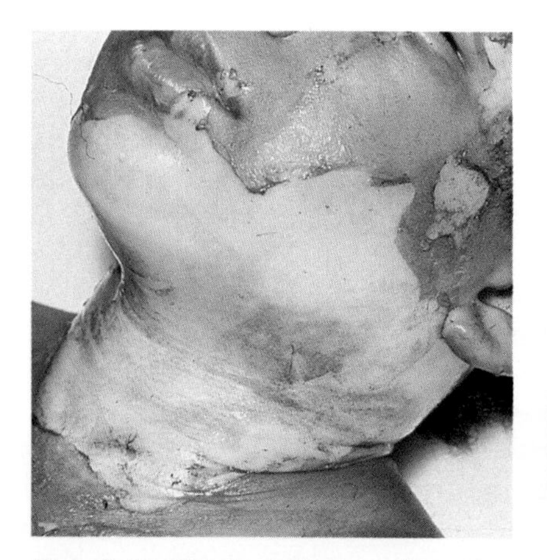

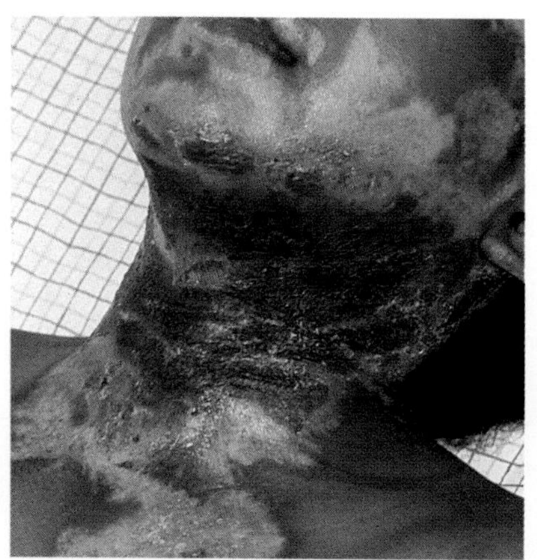

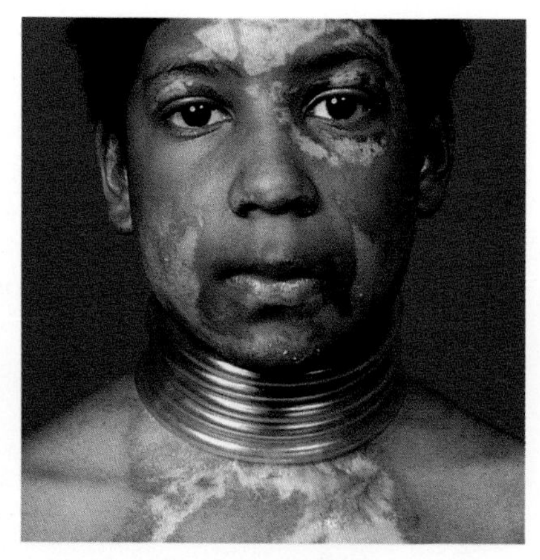

Fig. 6.12 Neck splints using thermoplastic suction tubing. (a) A deep burn to the neck.
(b) The burns have been excised and grafted.
(c) Splints are applied.

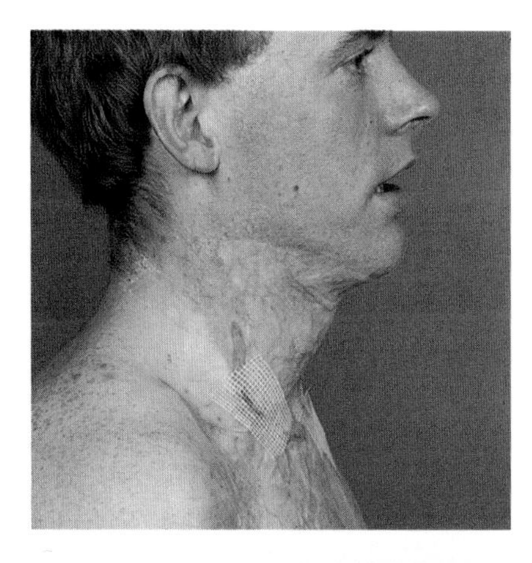

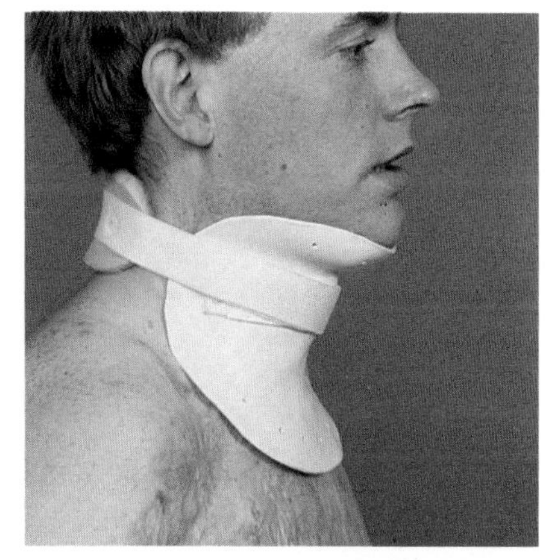

Fig. 6.13 Thermoplastic splint. (a) The wounds have been excised and grafted.
(b) The splint is applied and is worn constantly for many months.

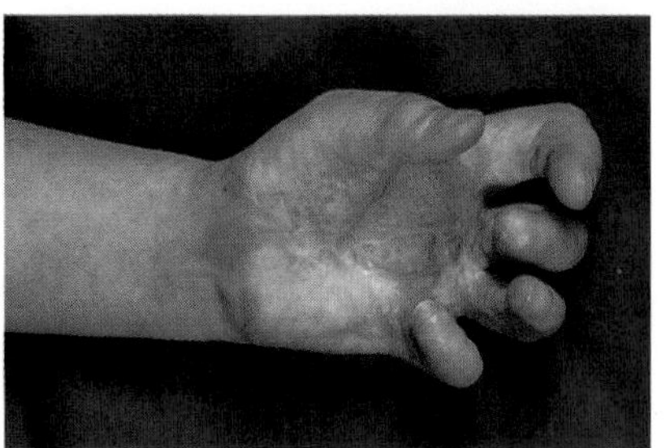

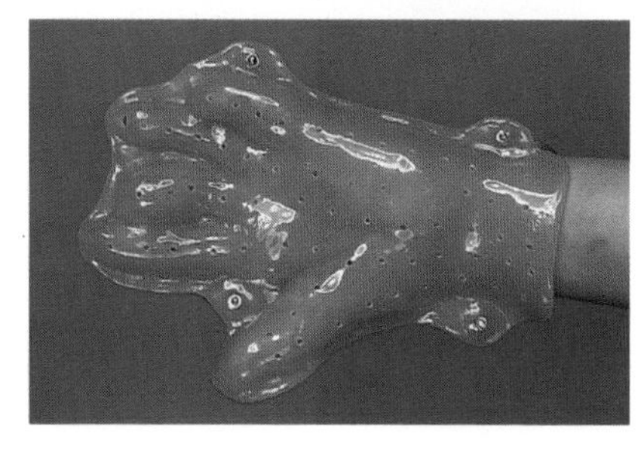

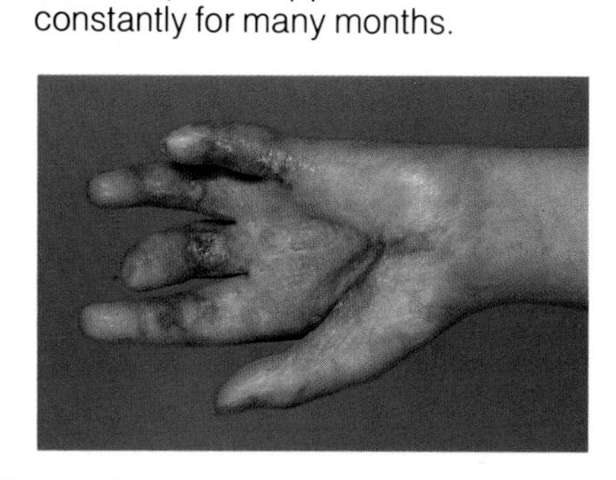

Fig. 6.14 Immobilization of the hand. (a) Burn wound contractures following a deep burn to the hand. Correction is required to allow normal development.
(b) The scars have been excised, the defects filled with full thickness grafts, and the hand immobilized in a bi-valve splint.
(c) After three months.

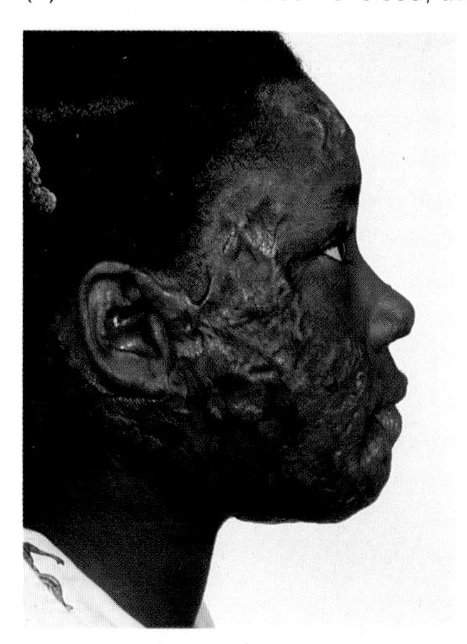

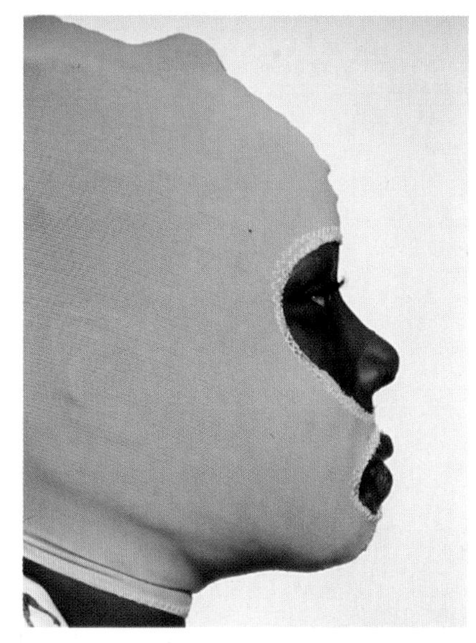

Fig. 6.15 Elasticated pressure garment. (a) These scars will gradually flatten out if this garment is worn.
(b) The Jobst mask.

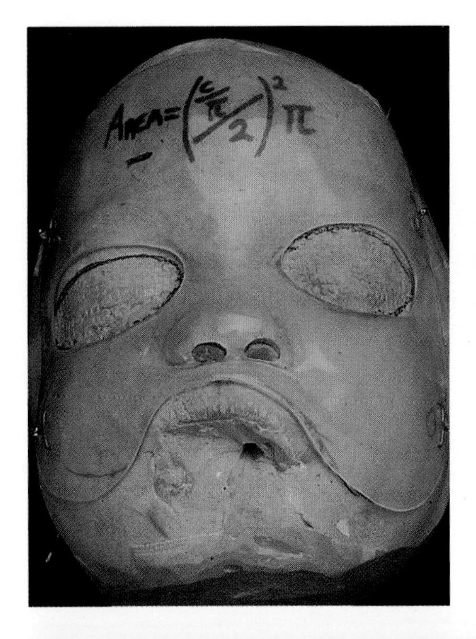

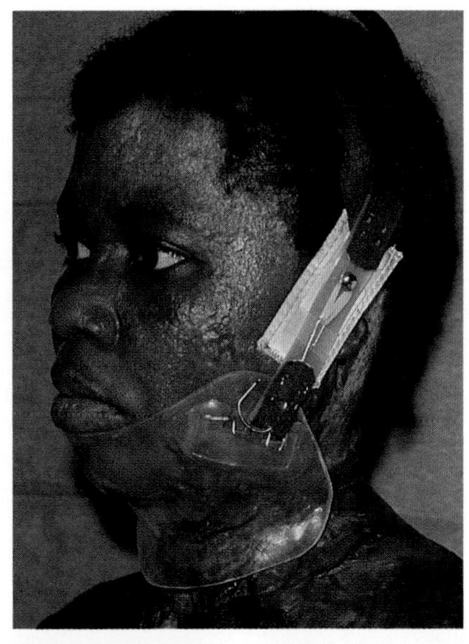

Fig. 6.16 Transparent flexible masks. (a) These masks are moulded on a plaster model. (b) Pressure can be applied to hypertrophic scars of the face. (c) They are held on with a simple orthodontic headpiece.

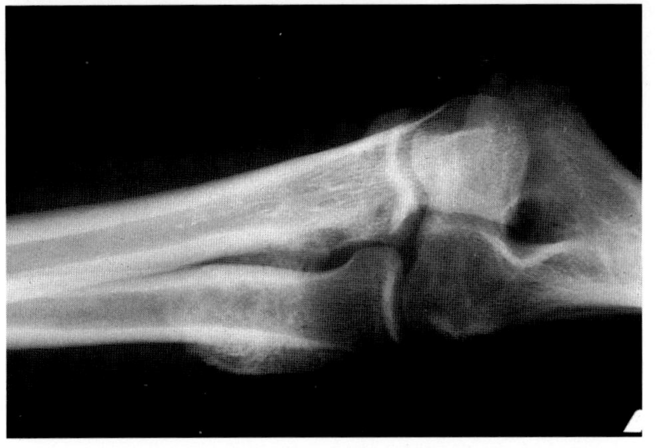

Fig. 6.17 Heterotopic bone deposition. This is seen at the biceps tuberosity of the radius and the medial epicondyle. It is mature and well trabeculated.

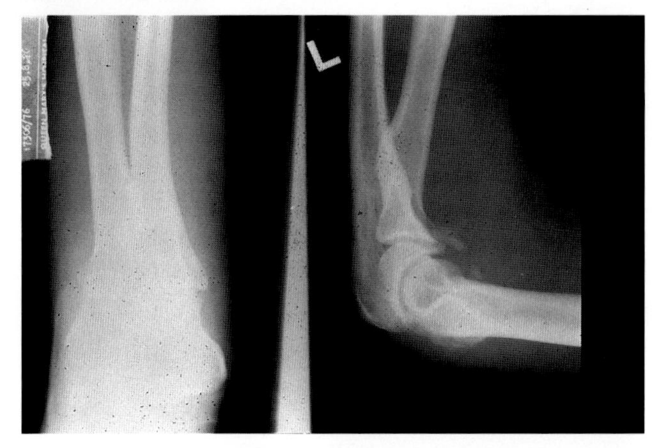

Fig. 6.18 Radiologically immature bone. Fluffy bone, not demonstrating trabeculae, has been deposited in the olecranon fossa and anteriorly.

HETEROTOPIC CALCIFICATION

This rare complication can occur around joints, even though a cutaneous burn may not be present in that area. Bone salts are deposited around the joint capsule (Fig. 6.17) and a gradual loss in the range of movement occurs; X-rays indicate calcified deposits, which become less fluffy and more discrete with time. Aetiology is unknown. Exercises which attempt to increase the range of movement only serve to accelerate the condition and must be avoided. Management is limited to preservation of strength and movement until calcification is mature. If trabeculation is indicated radiographically, the heterotopic bone which is blocking articulation can be excised and vigorous physiotherapy begun immediately. Attempting to remove the bone before it is radiologically mature encourages rapid recurrence of the condition (Fig. 6.18).

Calcification is seen most frequently around the elbow (Fig. 6.19). The olecranon fossa becomes filled with bone, limiting extension, and a build-up of bone around the biceps tuberosity, limits pronation and supination. Heterotopic calcification around the shoulder (Fig. 6.20), hip and knee is observed less frequently and is more difficult to treat.

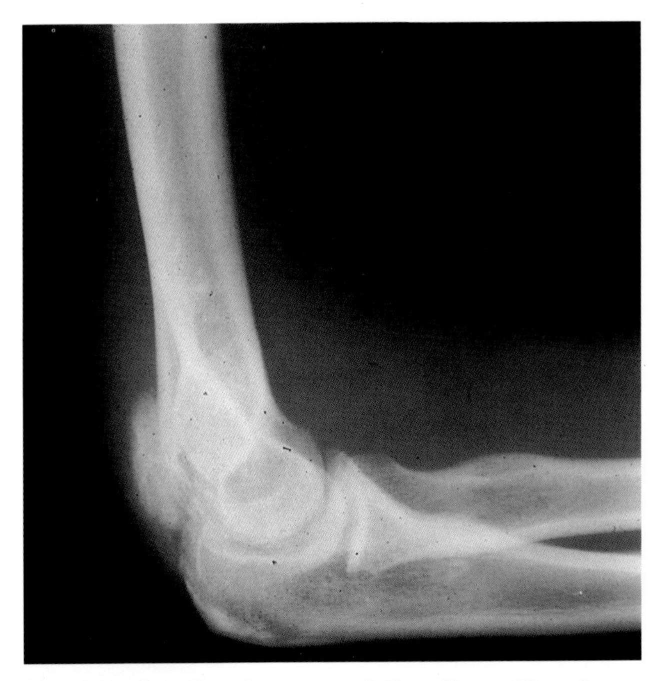

Fig. 6.19 Calcification around the elbow. The olecranon fossa has filled with heterotopic bone, preventing extension at the elbow.

Fig. 6.20 Heterotopic bone at the surgical neck of the humerus.

RENAL STONES

Historically, renal stones were seen in association with large burns, where there had been considerable weight loss, slow healing and a delay in mobilization. Renal stones are now very rare and confined to cases with large burns, with fractures of the long bones and chronic urinary infections. The presence of calculi is not usually recognized until the patient assumes a vertical position and develops acute renal colic. The stones are very small and repeated attacks of painful colic may be avoided by inserting ureteric stents; these allow free drainage of urine but deny passage of the stones. Definitive treatment by ultrasound can then await general clinical improvement.

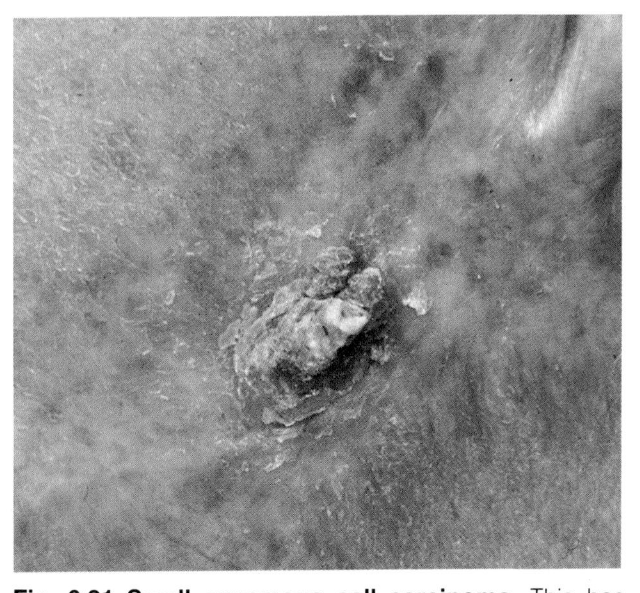

Fig. 6.21 Small squamous cell carcinoma. This has arisen in a 20 year old burn scar.

PIGMENTATION AND MALIGNANT CHANGES

Malignant changes can occur in older burn scars and usually form many decades after the injury (Fig. 6.21), particularly in areas that were not grafted and are subject to trauma and irritation. The malignant potential of these squamous cell cancers is variable, but they frequently metastasize rapidly after their excision.

Hypopigmentation can arise following deep dermal burns (Fig. 6.23). Management is by shaving the area and overgrafting with a very thin graft.

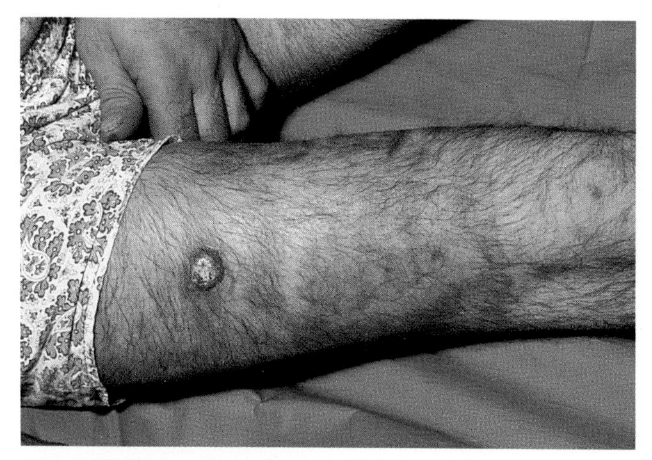

Fig. 6.22 Keratoacanthoma. This has arisen in a recent donor site on the upper thigh.

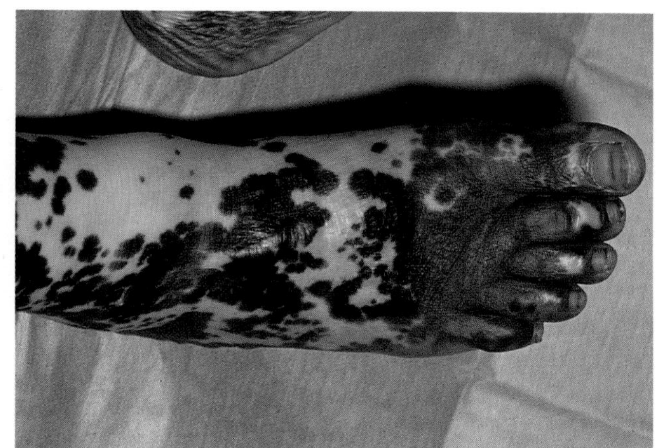

Fig. 6.23 Marked hypopigmentation. (a) This has followed a deep dermal burn.

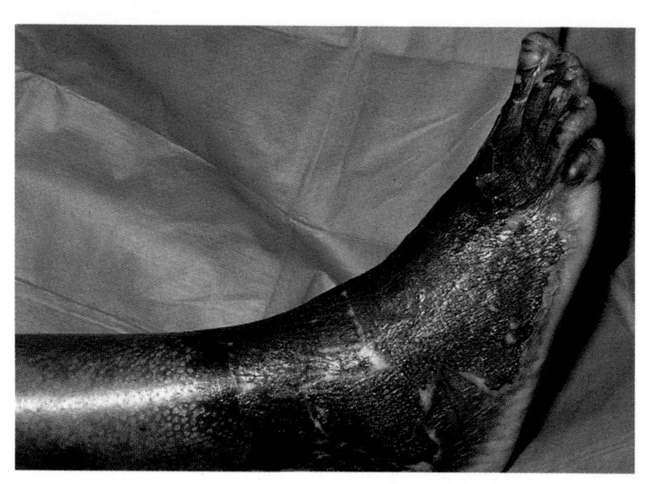

(b) The area has been shaved down and over grafted with a very thin skin graft.

NEGLECTED BURN INJURIES

This type of injury arises from failure to provide adequate care, usually for significant medical reasons or due to difficulties in arranging transportation from another country, rather than deliberate neglect or incompetence. The presence of the burn may have been denied by the patient or relatives, and medical advice avoided. Alternatively, the full significance of the injury may not be appreciated by medical attendants, and help or appropriate treatment is not sought for several weeks (Fig. 6.24).

The patient is often malnourished, with unclean wounds which are heavily colonized with organisms from the bowel. After a thorough evaluation, treatment is directed towards cleaning the wound, improving nutrition and correcting anaemia. Overgranulations are treated with topical hydrocortisone

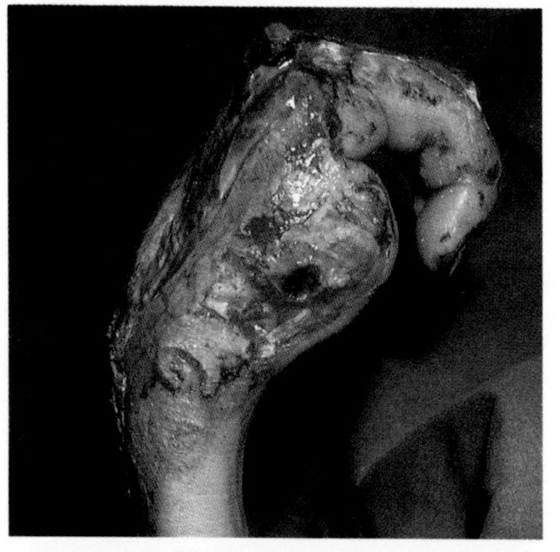

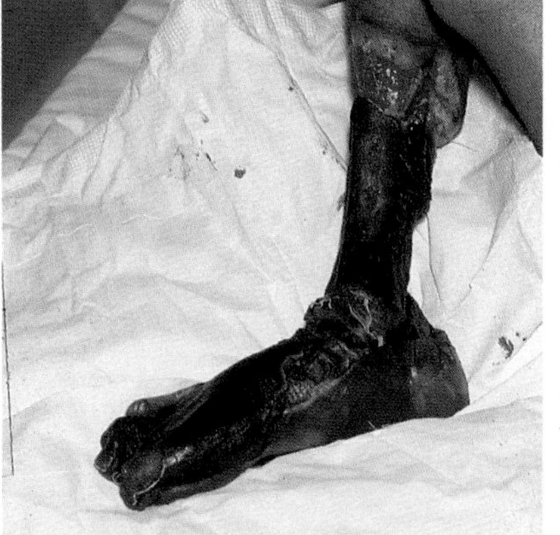

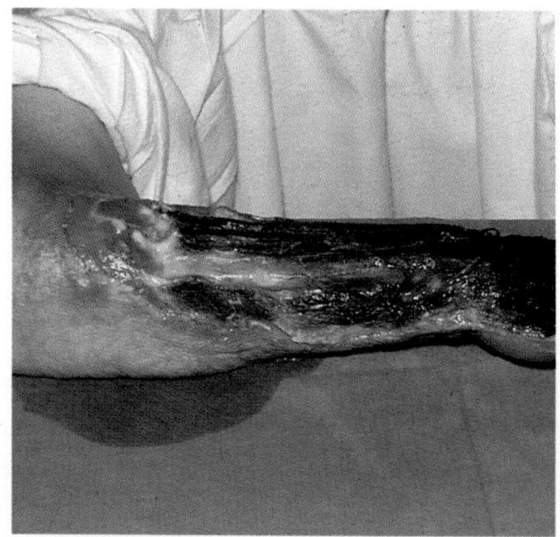

Fig. 6.24 Neglected burns. (a) Seven months after injury, skin grafts have been used to cover some of the raw areas in these burns of the hand but no attempt has been made to maintain function.
(b) A circumferential burn that did not receive an escharotomy. Total infarction and mummification of the foot is seen, which will require amputation.
(c) Delay in burn wound excision. A very deep burn on the extensor aspect of the arm. All the black structures needed excision some time ago, when there was some chance of preserving function.

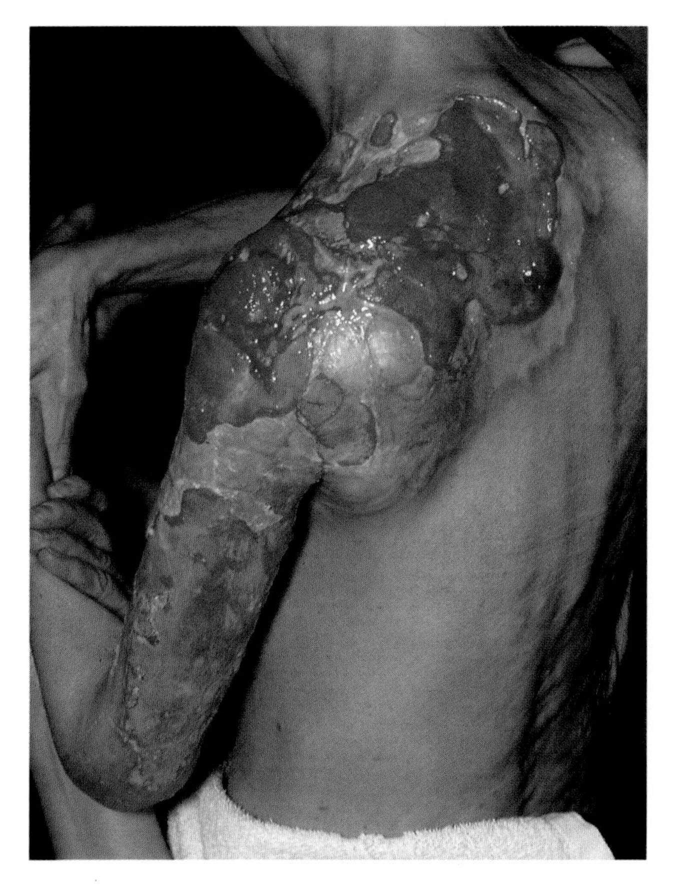

Fig. 6.25 Chronic over-granulation. Skin grafts have failed to become established on several occasions.

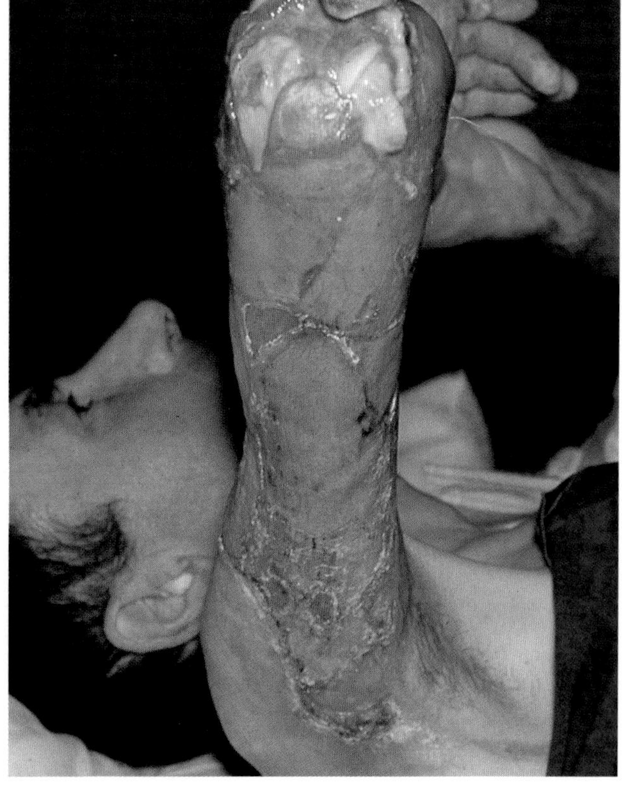

Fig. 6.26 Inadequate grafting. Skin grafts have been applied to the granulating areas; however, the elbow joint remains exposed. Excision of the non-vital structures is required, followed by flap cover, using either a free flap or jump flap from the abdominal wall.

ointment, and the slough is sharply excised (Fig. 6.25). A skin graft will then take well. In most cases, only one anaesthetic is needed to prepare the wound. When conditions are optimal a thin split skin graft can be applied (Fig. 6.26).

NON-ACCIDENTAL INJURIES

Non-accidental injury includes the deliberate infliction of burns. From a clinical and social aspect, cases involving children are the most important. The pattern of burning may raise doubts immediately (Fig. 6.29), but more often it is the inconsistencies in the history which arouse suspicion: different versions of the accident may be recounted, and there is often a delay of many hours before help is sought. The burn injuries often extend to a small area (Fig. 6.27) and frequently involve the buttocks (Fig. 6.28). There is a much greater incidence in boys compared with girls, and half of the children have been injured previously, with approximately one-quarter already on the 'at risk' register. Statistically, over half of the patients belong to families of a lower socio-economic status and one-third from single parent families.

In these difficult problems, treatment of the child is paramount: prompt admission to a 'place of safety' is mandatory. A thorough physical examination is important, with X-rays if indicated, and immediate help must be sought from the appropriate paediatrician and social worker, who will take responsibility for co-ordinating discussions and meetings. It is important that the burn physician and nurses remain on friendly and non-hostile terms with the family of the patient throughout the hospital stay, and ensure that long-term medical follow-up is adequate.

In the adult, non-accidental burn injuries are more likely to be seen in victims of assault (Fig. 6.30), those in institutionalized care (Fig. 6.31), or those suffering from psychosis, where the injuries may be self inflicted (Fig. 6.32). Cases of assault usually result in criminal proceedings and it is recommended that

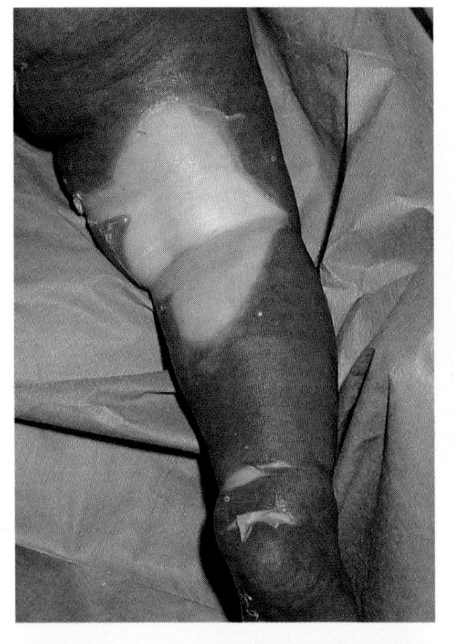

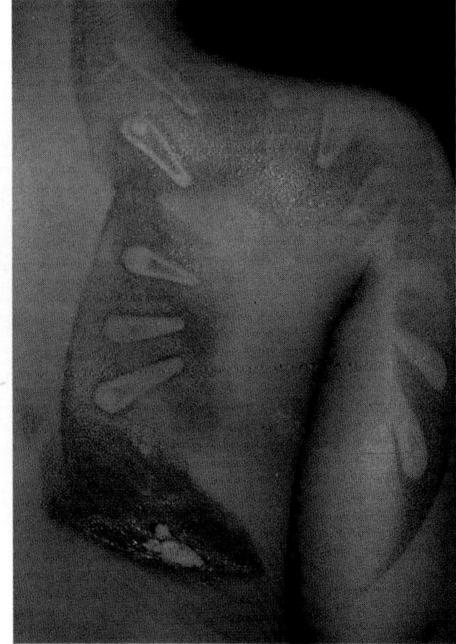

Fig. 6.27 Non-accidental injuries. (a) Scald injuries in a small child. The legs were obviously tightly flexed and the foot hyperextended at the time of the accident.
(b) Marks from a hot iron are clearly seen.

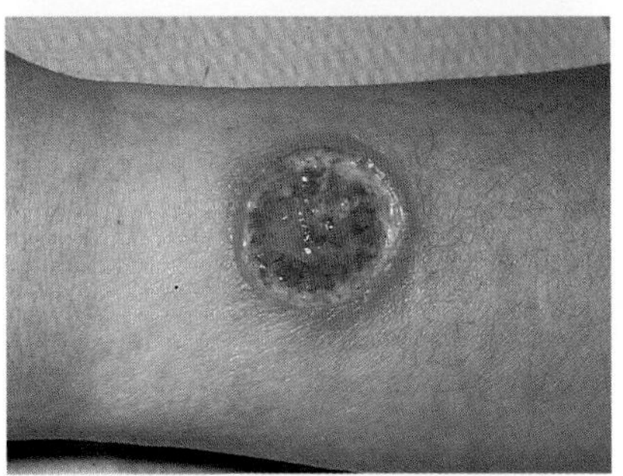

Fig. 6.28 A cigarette burn in a two year old.

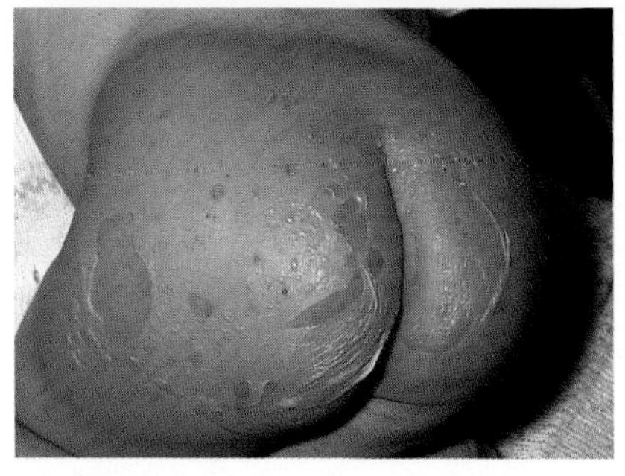

Fig. 6.29 Burns involving the buttocks. These are often associated with difficulties in toilet training.

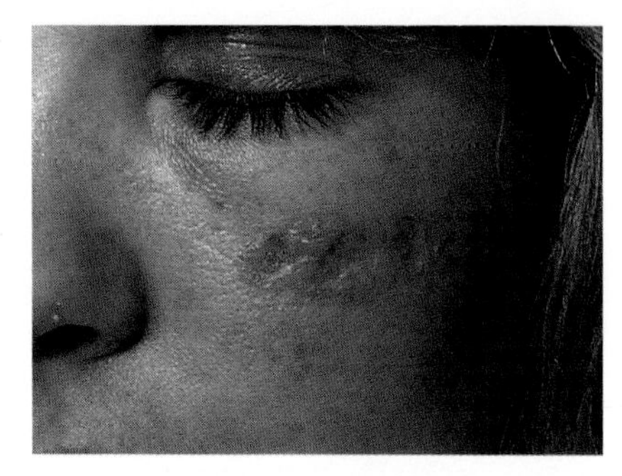

Fig. 6.30 Non-accidental burns in an adult. A row of cigarette burns seen in a victim of an assault.

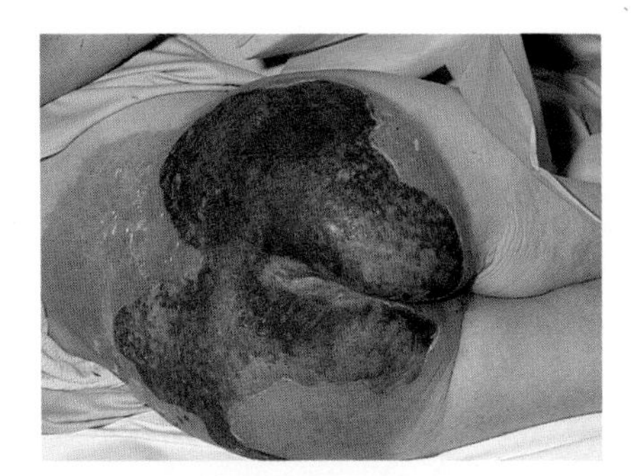

Fig. 6.31 Scald injuries. This patient, under care in a psychogeriatric unit had been placed in a bath that was too hot.

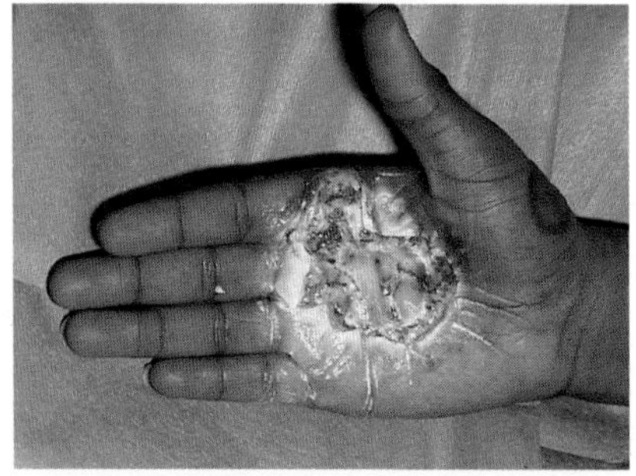

Fig. 6.32 Self-inflicted burn injuries. There was a similar 'stigma' on the other hand, which occurred in a schizophrenic patient during a phase of religious hysteria.

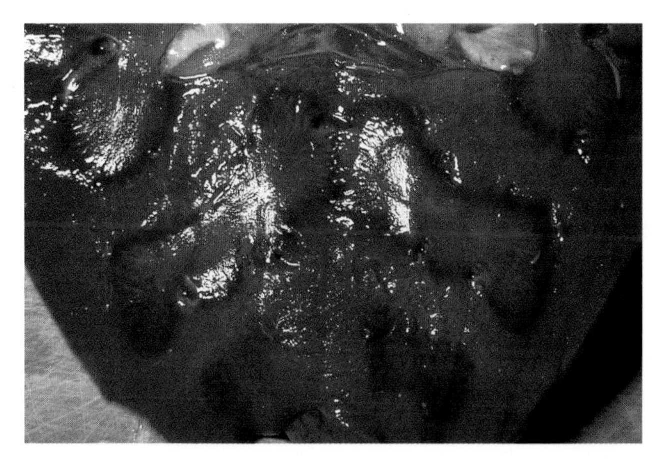

Fig. 6.33 Congested kidneys with vascular stasis in the renal medulla. Terminal renal failure followed uncontrolled sepsis.

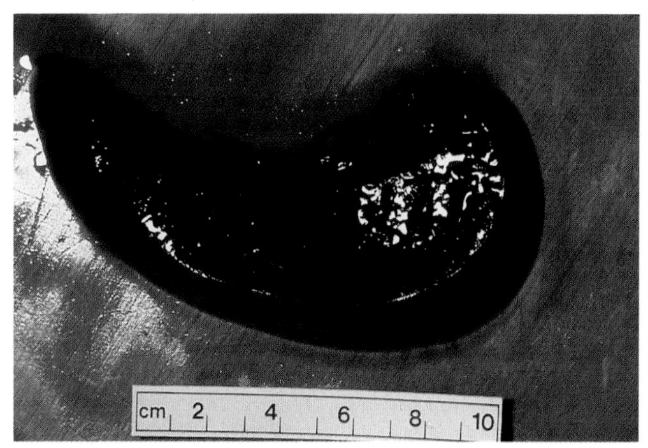

Fig. 6.34 Congested, soft and heavy spleen. The patient died from sepsis.

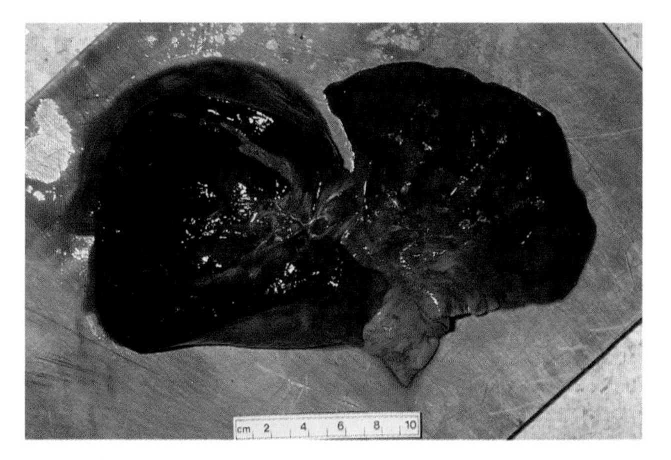

Fig. 6.35 Heavy consolidation of the lungs following inhalational injury. The lungs are filled with protein-rich oedema fluid and are relatively stiff and inelastic. PEEP is needed to inflate such lungs.

thorough notes and colour photography are taken by the medical attendants, to be used as subsequent evidence.

PATHOLOGY

It is remarkable how unusual it is to ascertain the exact cause of death at post-mortem examination in the majority of those patients dying from burns. Even those who have slowly fallen into uncontrolled sepsis and multiple organ failure rarely show more than congestion of most of the organs (Fig. 6.33), with enlargement and friability of the spleen (Fig. 6.34) and pulmonary consolidation (Fig. 6.35). Tissue autolysis occurs quickly and little help is available from histological examination. Death occurs at the biochemical level and gross derangements are often observed during the days preceding the patient's death.

Examination of the respiratory tract and lungs will pinpoint the focal damage following inhalational injuries (Fig. 6.36), but the changes due to the injury itself are obscured by the effects of secondary infection (Fig. 6.37).

Ante-mortem histological examination of the burn wound can be very useful in determining the presence of invasive burn wound sepsis, particularly those due to yeasts and fungi.

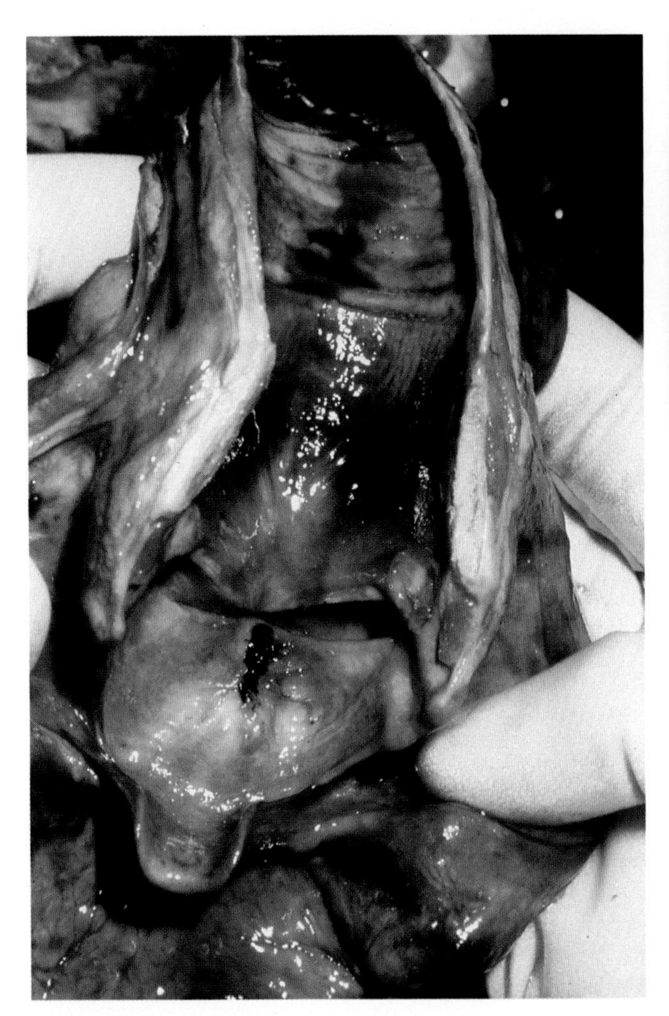

Fig. 6.36 Appearance of larynx and upper trachea from a case of inhalational injury. This shows smoke damage and early ulceration.

Fig. 6.37 Damage to the major bronchi following inhalational injury and secondary infection.

Further reading

GENERAL

Muir, I.F.K., Barclay, T.L. and Settle, J.A.D. (1987) *Burns and their management,* 3rd edition, Butterworth Heinemann, Oxford.
Wachtel, T.L. and Frank, D.H. (1984) *Burns of the head and neck,* W.P. Saunders, Philadelphia.
Hummel, R.P. (1982) *Clinical burn therapy,* John Wright, Boston.
Ruberg, R.L. (1986) Advances in burn care. *Clinics Plastic Surgery,* **13.1**, W.B. Saunders, Philadelphia.
Cason, J.S. (1981) *Treatment of Burns,*. Chapman & Hall, London.
Carvajal, H.F. and Parks, D.H. (1988) *Burns in children,* Year Book Medical Publishers, Chicago.

CHAPTER 1 PATHO-PHYSIOLOGY

Arturson, G. (1980) Pathophysiology of the burn wound. *Ann. Chir. Gynaecol.,* **69**, 178–83.
Hunt, T.K. (1990) Basic principles of wound healing. *J. Trauma,* **30**, (**Suppl.**)122–8.
Moritz, A.R. and Henriques, F.C. (1947) Studies of thermal injury: the relative importance of the time and surface temperatures in the causation of cutaneous burns. *Am. J. Pathol.* **23**, 695–720.

Area, Depth

Jackson, D.M. (1953) The diagnosis of the depth of burning. *Br. J. Surg.,* **40**, 588–96.
Lund, C.L. and Browder, N.L. (1944) The estimation of areas of burns. *Surg. Gynecol. Obstet.,* **79**, 352–4.

CHAPTER 2 CHEMICAL BURNS

Saydjari, R., Abston, S., Desai, M.H. and Herndon, D.N. (1986) Chemical burns. *J. Burn Care Rehabil.* **7**, 404–8.
Rozenbaum, D. and Baruchin, A.M. (1991) Chemical burns of the eye with special reference to alkali burns. *Burns,* **17**, 136–40.
James, N.K. and Moss, A.L.H. (1990) Review of burns caused by bitumen and the problem of its removal. *Burns,* **16**, 214–16.
Hunter, G.A. (1968) Chemical burns after contact with petrol. *Br. J. Plast. Surg.,* **21**, 337–41.
Wilson, G.R. and Davidson, P.M. (1985) Full thickness burns from ready mixed cement. *Burns,* **12**, 139–45.

Electrical burns, lightning

Bingham, H. (1986) Electrical Burns. *Clin. Plast. Surg.,* **13**, 75–85.
Hunt, J.L., Mason, A.D., Masterson, T.S. *et al.* (1976) The pathophysiology of acute electrical injuries. *J. Trauma,* **16**, 335–40.
Strasser, E.J. (1977) Lightning injuries. *J. Trauma,* **17**, 315–19.

CHAPTER 3 RESUSCITATION

Demling, R.H. (1987) Fluid replacement in burned patients. *Surg. Clin. North. Am.,* **67**, 15–22.
Settle, J.A.D. (1974) Urine output following severe burns. *Burns,* **1**, 23–42.
Carvajal, H.F. (1980) A physiological approach to fluid therapy in severely burned children. *Surg. Gynecol. Obstet.,* **150**, 379–84.

CHAPTER 4 INHALATIONAL INJURY

Kinsella, J. (1988) Smoke inhalation. *Burns,* **14**, 269–79.
Herndon, D.N., Barrow, R.E., Linares, H.A. *et al.* (1988) Inhalational injury in burned patients: effects and treatment. *Burns,* **14**, 349–56.
Boutros, A.R., Hoyt, J.L., Boyd, W.C. and Hartford, C.E. (1977) Algorithm for management of pulmonary complications in burn patients. *Crit. Care Med.,* **5**, 89–92.

Metabolism

Henley, M. (1989) Feed that burn. *Burns,* **15**, 351–61.
Wilmore, D.W. (1974) Nutrition and metabolism following thermal injury. *Clin. Plast. Surg.,* **1**, 603–19.
Curreri, P.W. (1975) Metabolic and nutritional aspects of thermal injury. *Burns,* **2**, 16–21.
Sutherland, A.B. (1976) Nitrogen balance and nutritional requirements in the burn patient: a reappraisal. *Burns,* **2**, 238–44.

Infection

Luterman, A., Dasco, C.C. and Curreri, P.W. (1986) Infection in burn patients. *Am. J. Med.,* **81 (1A)**, 45.
Lawrence, J.C. (1985) The bacteriology of burns. *J. Hosp. Infect.,* **6 (B)**, 3–17.
Winkelstein, A. (1989) What are the immunological alterations induced by burn injury? *J. Trauma,* **22**, 72–80.
Cole, R.P. and Shakespeare, P.G. (1990) Toxic shock syndrome in scalded children. *Burns,* **16**, 221–24.

CHAPTER 5 ANAESTHESIA

Park, R.E. (1982) Anaesthesia in the burn patient, in *Clinical Burn Therapy* (ed. P.D. Hummel), John Wright, Boston.
De Camp, T., Aldrete, J.A. (1981) The anaesthetic management of the severely burned patient. *Intens. Care Med.,* **7**, 55–62.

Wound excision and closure

Janzekovic, Z. (1970) A new concept in the excision and immediate grafting of burns. *J. Trauma,* **10**, 1103–8.
Hunt, J.L., Sato, R. and Baxter, C.R. (1979) Early tangential

excision and immediate mesh autografting of deep dermal hand burns. *Ann. Surg.*, **189**, 147–51.
Sykes, P.J. and Bailey, B.N. 1975) Treatment of hand burns with occlusive bags. *Burns*, **2**, 162–6.
Alexander, J.W., Macmillan, B.W., Law, E. and Kittur, D.S. (1981) Treatment of severe burns with widely expanded meshed skin autograft and meshed skin allograft overlay. *J. Trauma*, **21**, 433–41.
McMillan, B.G. (1978) Closing the burn wound. *Surg. Clin. North Am.*, **58**, 1205–31.
Burke, J.F., Bondoc, C.C. and Quinby, W.C. (1974) Primary burn excision and immediate grafting. *J. Trauma*, **14**, 389–95.
Deitch, E.A. and Staats, M. (1982) Child abuse through burning. *J. Burn Care Rehabil.*, **3**, 89–94.
Prasad, J.K., Thomson, P.D. and Feller, I. (1987) Gastrointestinal haemorrhage in burn patients. *Burns*, **13**, 194–7.

Physiotherapy

Parks, D.H., Evans, E.B. and Larson, D.L. (1978) Prevention and correction of deformity after severe burns. *Surg. Clin. North Am.*, **58**, 1279–89.
Ward, R.S. (1991) Pressure therapy for the control of hypertrophic scar formation after burn injury: a history and review. *J. Burn Care Rehabil.*, **12**, 257–62.
Edlich, R.F., Horowitz, J.H., Rheuban, K.S. *et al.* (1985) Heterotopic calcification and ossification in the burn patient. *J. Burn Care Rehabil.*, **6**, 363–71.

CHAPTER 6 MALIGNANT CHANGE

Bartle, E.J., Sun, J.H. and Wang, X.W. (1990) Cancers arising from burn scars. *J. Burn Care Rehabil.*, **11**, 46–9.
Fishman, J.R.A. and Parker, M.G. (1991) Malignancy and chronic wounds. *J. Burn Care Rehabil.*, **12**, 218–23.

Non accidental injury

Deitch, F.A. and Staats, M. (1982) Child abuse through burning. *J. Burn Care Rehabil.*, **3**, 89–94.
Lenoski, E.F. and Hunter, K.A. (1977) Specific patterns of inflicted burn injuries. *J. Trauma*, **17**, 842–6

Index

References in **bold** are to figures and references in *italic* are to tables.